Progress in Electro-Optics
Reviews of Recent Developments

NATO ADVANCED STUDY INSTITUTES SERIES

A series of edited volumes comprising multifaceted studies of contemporary scientific issues by some of the best scientific minds in the world, assembled in cooperation with NATO Scientific Affairs Division.

The series is published by an international board of publishers in conjunction with NATO Scientific Affairs Division

A	Life Sciences	Plenum Publishing Corporation
B	Physics	New York and London
C	Mathematical and Physical Sciences	D. Reidel Publishing Company Dordrecht and Boston
D	Behavioral and Social Sciences	Sijthoff International Publishing Company Leiden
E	Applied Sciences	Noordhoff International Publishing Leiden

Progress in Electro-Optics

Reviews of Recent Developments

Edited by

Ezio Camatini

Technology and Industrial Plant Division
Department of Mechanical and Machine Engineering
Polytechnic Institute of Milan
Milan, Italy

PLENUM PRESS • NEW YORK AND LONDON
Published in cooperation with NATO Scientific Affairs Division

Library of Congress Cataloging in Publication Data

Nato Advanced Study Institute on Progress in Some Areas of Electro-Optics: the
   Alphanumeric Displays, Milan, 1973.
   Progress in electro-optics.

   (NATO advanced study institutes series: Series B, Physics; v. 10)
   Includes bibliographical references and index.
   1. Electrooptical devices—Congresses. I. Camatini, Ezio. II. Title. III. Series.
TA1750.N37 1973   621.38'0414   75-16350

   ISBN 978-1-4684-2792-9   ISBN 978-1-4684-2790-5 (eBook)
   DOI 10.1007/978-1-4684-2790-5

Lectures presented at the 1973 NATO Advanced Study Institute on Progress
in Some Areas of Electro-Optics: The Alphanumeric Displays,
held in Milan, Italy, August 23-31, 1973

© 1975 Plenum Press, New York
Softcover reprint of the hardcover 1st edition 1975

A Division of Plenum Publishing Corporation
227 West 17th Street, New York, N.Y. 10011

United Kingdom edition published by Plenum Press, London
A Division of Plenum Publishing Company, Ltd.
Davis House (4th Floor), 8 Scrubs Lane, Harlesden, London, NW10 6SE, England

CONTRIBUTORS

H. J. Caulfield, Block Engineering, Inc., Cambridge, Massachusetts (U. S. A.)

C. H. Gooch, Ministry of Defence, Service Electronics Research Laboratory, Baldock, Hertfordshire (United Kingdom)

O. S. Heavens, University of York, Heslington, York (United Kingdom)

A. Kastler, Nobel Prize, Université de Paris, Ecole Normale Supérieure, Laboratoire de Physique, Paris (France)

W. E. Kock, Acting Director, The Herman Schneider Laboratory of Basic and Applied Sciences, University of Cincinnati, Cincinnati, Ohio (U.S.A.)

P. B. Page, ITT Components Group Europe, Central Applications Laboratory, Harlow, Essex (United Kingdom)

E. Ritter, Balzers AG, Balzers (Fürstentum Liechtenstein)

U. J. Schmidt, Philips Forschungslaboratorium, Hamburg (West Germany)

G. F. Weston, Mullard Research Laboratories, Redhill, Surrey (United Kingdom)

PREFACE

Advances in systems technology are creating the need for alphanumeric displays and component technology to satisfy this need. The field of alphanumeric displays covers applications from the single character lift indicator to the thousand-plus character computer readout. A survey of the state of alphanumeric displays helps the user of display devices to make a choice, for a particular application, between the various devices available now or in the near future. It is essential to consider the circuits and the display device together in order to obtain a clear picture of the economics of the different techniques.

In general, a display module is controlled by binary input signals at normal logic powers and may be subdivided into five basic elements: 1) data memory, 2) character generator, 3) driving circuits, 4) scanning circuits, and 5) display device.

The data memory is essential to make the display module independent of the system. Normally, this will be an electronic memory, but in some cases, the display device will have an inherent memory. The character generator must perform two functions: 1) convert the binary code to a '1 out of n' form to select the chosen character out of the 'n' available, and 2) create the character format, although in some cases this is inherent in the display device. The driving circuits are required to interface between the display device and the logic circuits, as in most cases either higher voltage or current or both are required to operate the display device. The scanning circuit controls the positioning of each character on the display device. When electronic memories are used, these are synchronized to the scanner to provide a continual refreshing of the displayed data at a rate high enough to minimize flicker. There are a wide variety of display devices using gas, phosphor, incandescent and semiconductor light sources. Some are fixed format devices and thus only need part 1) of the character generator above. Others are variable format and form the characters from bars, dots, etc.

TYPES OF APPLICATIONS

It is helpful to subdivide the applications of displays, as they have a significant effect on the types of devices used and the arrangement of the circuits. Three broad sectors may be defined as follows:

Sector A. Displays for Instrumentation and Automation. This covers systems such as digital voltmeters and frequency meters, data displays in process control systems, etc. As in most cases, measured variables of known range are being displayed and trained operators are involved; range selectors can be provided. Displays with a large number of digits are, therefore, not required. A typical instrument may have a six digit display giving a resolution of 0.0001% of the full scale reading. Currently, most instruments display numeric data, and the formed cathode indicator tube is the most popular device. As the cost of semiconductor 5 X 7 dot displays comes down, new applications will emerge using alphanumerics. The small size of monolithic Ga(AsP) displays also creates new possibilities. An example of the combination of the two above points is the "programmable key" concept. By closely associating a group of arrays with a push-button key, the function and label of the key may be computer controlled. The scope for simplifying or extending to range of control consoles by using this technique is enormous.

Sector B. Displays for Calculating and Similar Machines. This includes data handling displays of similar size. Use, in general, by non-technical personnel, the wide range of calculations and the need of speed preclude the use of range switching. Typical machines display a single row of 16 characters. As with Sector A, currently, most machines display numeric data, and the formed cathode indicator tube is the most popular device. The trend towards programmed calculating machines and other such devices is, however, creating a demand for alphanumeric displays in this field. An example of a new application would be a display for the visual checking of a line of keyboard data before entering it into the data processing computer.

Sector C. Displays for Computer Readout. This includes data processing applications such as the display of stock market figures, details of bank accounts, flight information at airports, etc., where large amounts of information must be displayed often to groups of viewers. A typical arrangement may display 800 characters and would have a keyboard and controls so that data could be typed, sent or received. Hence, the trend for some time has been to alphanumeric displays and the cathode ray tube is by far the most common device.

TYPES OF DISPLAY DEVICES

The following summarizes briefly some of the more important types of display devices that are produced or are in advanced stages of development. Numerical Indicator Tubes (NIT). The formed wire cathode NIT is a very well-known cold cathode, gas filled, indicator tube and is widely used in Sectors A and B defined above. It offers good character shape, but its major disadvantage is that the number of cathodes that can be stacked is limited to about ten; thus, the device is normally only used to display numerical data. There is also a peak current rating that limits the size of a time shared group to about 16 characters. Bar Tubes (BAR). Devices using the bar format can be constructed using many techniques, e.g., gas discharge tubes, vacuum fluorescent tubes, incandescent bars, etc. Compared with the NIT character, shape is inferior, but the characters are in one plane. Character generation is dearer than that for the NIT but cheaper than for a 5 X 7 dot display. Semiconductor Light Emitting Diodes (LED). Currently, the most commonly used semiconductor materials are gallium arsenide phosphide (Ga (AsP)) and gallium phosphide (GaP). Arrays are normally formed in a seven-bar format for numeric and a 5 X 7 dot format for alphanumeric displays. These devices are diodes that emit narrow spectrum light when forward biased. The array formats can be produced from discrete diodes mounted on a substrate; but with (Ga (AsP)), it is also possible to produce a monolithic structure. With this monolithic technique, 5 X 7 dot displays of only 3 mm height (slightly larger than a typewritten capital letter) have been made, and smaller sizes are practical. LED will compete in sections A and B, although high cost initially will restrict their use to applications where small size, high reliability and long life are important. Mullard (Philips) Dot Gas Tube (DGT). This device is a derivation of the NIT and consists of a common anode and 35 short wire cathodes arranged in a 5 × 7 dot matrix. Production costs of this device should be similar to that for the NIT, but its big advantage is that it can display a wide range of alphanumeric and special characters in a single plane with high brightness and a wide viewing angle. Compared with the NIT, the main disadvantage for numeric only applications is that extra electronics are required to encode from decimal into the 5 × 7 patterns. Also, as with the NIT, time-shared groups are limited to about 16 characters. This device will compete in sectors A and B, particularly where alphanumeric data are to be displayed. Burroughs Self-Scan (S-S). Like the DGT, this device displays characters in a 5 × 7 dot format; it is, however, fabricated as a single unit multi-character display. For 16 characters, a unit is 7 dots high and 112 dots wide (two blank columns between each 5 column character). The design is unique in that scanning is simplified to a three-phase plus reset signal that is addressed to the cathodes and that "scans" a column of 7 dots along the back of the display between the cathodes and "back" anodes. Data are presented a column at a time to 7 "front" anodes, which effectively run the length of the display, and the glow is pulled through to the front as required to achieve the character dot pattern. The S-S is aimed at sector B and the lower end of C for display of alphanumeric data. Disadvantages are the same as for the DGT. Plasma Panel. This device consists essentially of two sheets of glass with a gas filled space in between; on the outside of the sheets are deposited transparent conductors in an X − Y (orthogonal) format. The device is AC operated at a few hundred kilohertz and has an inherent memory. Because of the memory, peak current and refresh rate offer no problem; thus, for large arrays, this is superior to devices like the S-S. Although the display element is simple and should be cheap to make, the associated electronics are complex for small groups of characters. Thus, the sectors will be in the upper end of sector B and throughout sector C. The P-P is in early stage of its development, so many advantages and disadvantages are unknown. It does, however, have considerable potential as a competitor to the CRT, as colour and multilayer displays are feasible. Cathode Ray Tube (CRT). Currently, this is the cheapest method of achieving large scale displays for sector C. However, CRT's without memory require expensive associated electronics to refresh the display above flicker rate; CRT's without memory have a short life. Attempts have been made to produce CRT's for sectors A and B, but modules using these are normally more expensive than other techniques. Because of the problems involved with these devices, much effort is being made to find an alternative. The Plasma-Panel, so far, seems the most interesting approach to this problem.

CIRCUIT TECHNIQUES

The basic fundamentals of display circuits are the following ones. Static Displays. "Static" is used to mean that all the display devices are on together; thus, each requires its own character generator. Where the output of a counter is to be displayed, it may be possible to omit the memory. For numeric displays using NIT and some BAR devices, standard integrated circuits exist that combine the function of character generation and driving in one package and are simpler than the alphanumeric character generator described above. Time-shared Displays. "Time-shared" is used to mean that one character generator is shared between a group of display devices. This means that only one display device is on at any one time and that presentation of the data from the memory and the switching of the scanner are synchronized; the method does have the disadvantage that in order to maintain the same brightness compared to the static displays, the current flowing in the display device when it is on must be higher than the normal value DC. With 5×7 dot display devices capable of $X - Y$ address, however, it is possible to use two alternative scanning arrangements to advantage. <u>Multiline Displays.</u> If the display device does not have inherent memory, the time-shared groups in a multiline display must be scanned simultaneously, and for a large unit there may be more than one group to a line. Simultaneous scanning means that time-sharing of the character generator between lines becomes more complex and expensive. Even with a display device having inherent memory, it is essential that addressing of the points is in an $X - Y$ manner in order to reduce the cost of the writing circuits. The Plasma-Panel meets these requirements. For CRT displays, there are many circuit variants, but basically they fall into three broad classes: 1) shaped beam, which creates the character shape by passing the beam through a mask; 2) selective scanning, which creates each character individually by curves, strokes or dots; 3) raster scanning, which builds up the whole group of characters from dots, as with TV. High fixed costs tend to limit the use of 1) and 2) to large displays of, say, 500-1000 characters. Method 3), however, although less efficient, does enable displays of a few hundred characters to be made cheaply.

DISPLAY MODULE COSTS

In estimating the cost of a particular display application, it is important to include the relevant electronic costs, and not just the cost of the display devices. In fact, in most cases, the cost of the electronics will be more than 80% of the total cost. Thus, using a cheaper display device may not be the best solution if, in fact, it requires more expensive circuits; e.g., using the above proportion, a 20% reduction in device cost is cancelled by a 5% increase in electronic costs. With applications in section A, the electronic costs are dependent on whether a static or time-shared mode of operation is most relevant. The choice is dependent on whether the display is numeric or alphanumeric, as the character generator is much more expensive for the latter and, thus, the benefits of time-sharing are greater. With applications in sector B, the number of digits normally makes time-sharing most popular, even for numeric-only displays. It must be remembered that the display device's peak current limitation restricts the number of characters in a time-shared group. Thus, for large strips of characters, more than one group may be essential, and this will increase the fixed cost of the display module. In sector C, the large number of characters makes time-sharing essential. Refreshing the display above flicker rate is a major part of the cost, hence the interest in devices with inherent memory, such as the Plasma-Panel. An $X - Y$ address system is essential to minimize the cost of the driving and scanning circuits. If a wide variety of display problems are being considered, it is probably best to determine the coefficients of the formula: $C = F + Hx + Vy$, where C is total cost, F is fixed cost, H is cost per character in the horizontal direction, V is cost per character in the vertical direction, x is the number of characters in the horizontal direction, and y is the number of characters in the vertical direction. But care must be taken to consider points like those raised above, which can place limits on the range of x and y for which the coefficients are valid.

FUTURE TRENDS

Consider the display devices currently in various stages of research, it is possible to estimate the trend over, say, the next 5 to 10 years. In sector A, the availibility of low cost, solid-state alphanumeric display modules with mechanical and electrical compatibility with integrated circuits will open up a whole new range of applications. The "programmable" label" concept discussed is a forerunner of such application. The movement of automation into the consumer fields will create the need for clear display of alphanumeric information. The remarks made for sector A are similar to those for sector B. Such devices as the pocket "electronic slide-rule" become feasible. The field of data processing is wide open to create new applications once display module prices are low enough. For sector C, the situation is different, as large multi-row displays make a display device with inherent memory much more important; flat displays will also create new possibilities. The Plasma-Panel meets many of the requirements but needs a high-voltage high-frequency supply. Liquid crystals look interesting but are still at a very early stage in their development. With new techniques being considered, multi-character semiconductor displays seem a viable proposition. With most of the new devices, the ability to produce displays of various colours is under consideration, and solutions to the problem should occur in the near future. This is particularly interesting to sector C, if it makes flat, low cost colour displays practical. For more than a decade, the NIT and the CRT have dominated the data display field. Over the next decade, however, the rate of technological change in the display field will be far more rapid.

In the above connection, the Advanced Study Institute on "Progress in Some Areas of Electro-Optics: The Alphanumeric Displays" held in Milan (Italy) from August 23 to 31, 1973 under the sponsorship of NATO, Scientific Affairs Division, has proved to be very fruitful as a stimulating exchange of information between physicists and engineers in the state-of-the-art-and-technology in the subject of the meeting. The proceedings of the Institute reflect the level and the interest of the meeting and can be considered a powerful tool for all people working in the field of the alphanumeric displays and a valuable contribution to a greater knowledge and to the development of this progressing area of electro-optics.

The material in these proceedings aims to a focalization of the main topics, as follows: fundamentals of electro-optics; introduction to the alphanumeric displays; display devices; circuit techniques; applications: potential and present; display module costs; economics; future trends.

As scientific director of the Advanced Study Institute on "Progress in Some Areas of Electro-Optics: The Alphanumeric Displays" and editor of these proceedings, my gratitude goes to the Scientific Affairs Division of NATO, which sponsored the course and made possible the publication of this work. I also wish to thank the lecturers for their papers, which made this book a valuable document of the Institute.

Ezio Camatini

Politecnico di Milano
Italy

CONTENTS

THE HISTORICAL DEVELOPMENT OF ELECTRO-OPTICS

A. Kastler

Université de Paris, Ecole Normale Superieure, Laboratoire de Physique

Paris (France)

The historical development of electro-optics begins with the development of electromagnetism (5).

Already in the eighteenth century, in the year 1758, the German physicist Wilcke recognized that dielectrics introduced into an electric field become polarized, but it was only a century later, in 1837, that Faraday recognized the importance of dielectrics for electrostatic phenomena. His ideas inspired Maxwell in 1862 to introduce the concept of "displacement current." In the meantime, after the discoveries of Galvani and Volta in Italy, electromagnetism had been developed by Ampere and Faraday, and in 1852 Wilhelm Weber measured the ratio of electromagnetic and electrostatic units, having the dimension of a velocity, and found the surprising result that it was identical to the velocity of light.

This remarkable result showed that a connection must exist between light and electric phenomena. It led Maxwell, as you know, to the development of the electromagnetic theory of light, in the year 1865. As a consequence of this theory it became obvious that the phenomena of refraction and dispersion, and also of scattering of light by dielectric media, were produced by the dielectric polarization of the medium under the influence of the electric field of the electromagnetic wave. This was especially shown by the relation between the index of refraction, which is a function of the light frequency, and the electric permittivity ϵ of the medium in electrostatic fields, a relation which has the simple form

$$n^2 \ (\nu \to o) = \epsilon.$$

Thus, the basis for the understanding of electro-optics was ready, and it is not surprising that a few years after Maxwell's work, in 1875, the electric birefringence of dielectric media was discovered by Kerr and called, after its discoverer, the Kerr effect.

The first experiments were made in liquids, especially in CS_2 and nitrobenzene. Liquids are optically isotropic. If an electric field is applied to a liquid in the z-direction, it acquires the property of a uniaxial crystal with its optical axis in the direction of the applied field.

Kerr found that

$$n_e - n_o = B \ \lambda \ E^2$$

a quadratic effect.

λ is the wavelength of the light; B is called the Kerr constant; the phase retardation φ is given by

$$\frac{\varphi}{2\pi} = \frac{\delta}{\lambda} = \frac{(n_e - n_o) \ \ell}{\lambda} = B \ \ell \ E^2$$

where ℓ is the thickness of the medium; B is temperature-dependent and becomes smaller if the temperature is raised.

There is a dispersion of the Kerr constant given by Havelock's formula:

$$\frac{B \ \lambda \ n}{(n^2\text{-}1)^2} = C^{te}$$

After Kerr's discovery the effect of an electric field on the optical properties of crystals was investigated.

Crystals are naturally anisotropic and birefringent. It was expected that this birefringence could be modified by an electric field applied to a crystal plate. The first positive results were obtained in 1883 by Roentgen (7) (the physicist who later discovered X-rays) and by Kundt (8) (known for his work on acoustics). They both investigated quartz crystals.

Two years before, in 1881, Pierre and Jacques Curie (10) had discovered the property of piezoelectricity of quartz: that is, the fact that a mechanical stress in a given direction produces electric charges, as a consequence of an electric polarization of the crystal plate by the stress.

These two effects, piezoelectricity and electric-field-induced birefringence, are intimately connected; both can occur only in crystals without a center of symmetry.

We have seen that the birefringence of liquids produced by an applied electric field is a quadratic effect; it is proportional to E^2, the square of the applied field, and remains the same when the direction of the field is reversed. In constrast to this, the birefringence produced by applying a field to a centro-asymmetric crystal is a linear effect and changes sign when the field vector is applied in the opposite direction. It is obvious that for symmetry reasons, such an effect cannot exist in crystals having a center of symmetry.

The theory of this linear electro-optic effect was worked out by the German physicist Pockels (9) in 1893. For this reason the effect is often called the "Pockels effect", to distinguish it from the quadratic Kerr effect.

The optical properties of a crystal can be described by a second-order tensor which can be geometrically represented by the "index ellipsoid" introduced by Fresnel. The equation of this ellipsoid in a general coordinate system X, Y, Z can be written

$$\frac{X^2}{n^2_{11}} + \frac{Y^2}{n^2_{22}} + \frac{Z^2}{n^2_{33}} + \frac{2XY}{n^2_{12}} + \frac{2YZ}{n^2_{23}} + \frac{2ZX}{n^2_{13}} = 1$$

Such a tensor is defined by 6 parameters: "components" forming a matrix of "indices". Instead of using two indices to define each of these six parameters we can use one index only, making the following convention:

$$n_{11} = n_1 \; ; \; n_{22} = n_2 \; ; \; n_{33} = n_3 \; ; \; n_{23} = n_4 \; ; \; n_{13} = n_5 \; ; \; n_{12} = n_6$$

6 components
$$\begin{vmatrix} n_{11} & n_{12} & n_{13} \\ & (6) & (5) \\ & n_{22} & n_{23} \\ & & (4) \\ & & n_{33} \end{vmatrix}$$

This matrix can be diagonalized, which means that the equation of the ellipsoid is referred to its principal axes, defining the principal indices

$$n_1{}^o, \; n_2{}^o, \; n_3{}^o.$$

If we apply an electric field to the crystal in an arbitrary direction, in the general frame X, Y, Z, this field vector $\vec{E}$ is defined by 3 components E_x, E_y, E_z.

Each of these components E_i will modify each of the six tensor components:

$$\Delta \left(\frac{1}{n^2_j}\right) = \alpha_{ij} \, E_i \qquad (3 \times 6 = 18)$$

Thus, in the most general case, the effect of the field on the crystal is described by 18 parameters α_{ij} ranging from α_{11} to α_{63}, which are called electro-optical constants. Mathematically this means that the application of a 3-component vector to a 6-component tensor (2nd-order tensor) gives rise to a 18-component tensor (3rd-order tensor) whose components can be written α_{lki}. Geometrically the electric field applied in any direction deforms the index ellipsoid, changing the lengths and the orientation of each of its principal axes, and this change depends on the direction of the field.

18 parameters are necessary in the most general case when no symmetry element is present. This is the case of the triclinic crystal class. Symmetry elements reduce drastically the number of independent parameters. Take quartz as an example. Its behavior is described by two electro-optical constants α_{41} and α_{63}, if we choose for the optical axis of the uniaxial crystal the Z-axis (Z-axis: number 3). The existence of α_{63} means that if we apply an electric field in the 3-direction (E_Z parallel to optical axis), we create in the equation of the index ellipsoid a n_6 or n_{12} coefficient of the term X Y.

In the absence of the electric field the equation is

$$\frac{X^2 + Y^2}{n_o^2} + \frac{Z^2}{n_e^2} = 1$$

with the field, it becomes

$$\frac{X^2 + Y^2}{n_o^2} + \frac{Z^2}{n_e^2} + 2\,\alpha_{63}\ X\,Y = 1$$

rotating the axis in the XY plane by 45°

$$X'^2 \left[\frac{1}{n_o^2} + \alpha_{63}\,E_Z \right] + Y'^2 \left[\frac{1}{n_o^2} - \alpha_{63}\,E_Z \right] + \frac{Z^2}{n_e^2} = 1$$

$$n'_x = n_o - \frac{n_o^3}{2}\,\alpha_{63}\,E_Z$$

$$n'_y = n_o - \frac{n_o^3}{2}\,\alpha_{63}\,E_Z$$

Birefringence: $n'_y - n'_x = n_o^3\,\alpha_{63}\,E_Z$

During the nineteenth century, the linear electric birefringence effect was found on tourmaline, sodium chlorate (cubic system), and K-Na tartrate salt studied by Pockels. The magnitude of the Pockels effect is as follows: for quartz and similar crystals

$$n_o^3\,\alpha_{ij} \sim 10^{-6}$$

if we apply a field of $E = 10^4$ volts/cm:

$$\Delta n \sim 10^{-6} \qquad\qquad \ell = 5 \text{ cm}$$

$$\lambda = 0,5\,\mu = 0,5.10^{-4} \text{ cm}$$

$$\frac{\ell\,\Delta n}{\lambda} = \frac{5.10^{-6}}{0,5.10^{-4}} = 10^{-1} \qquad \varphi = \frac{2\pi}{10} = 36°$$

this birefringence is easily observable, but it needs high fields to be studied. For the K-Na tartrate salt Pockels found a birefringence 20 times that of quartz, but he found also that in this case a quadratic Kerr effect is superposed on the linear effect. The K-Na tartrate salt was the first case of a so-called "ferroelectric crystal", a crystal which below a Curie point temperature acquires a permanent electric moment. Above the Curie point temperature it is "paraelectric", but its polarizability is high and temperature-dependent. Since that time other ferroelectric crystals have been investigated and are of great practical importance, such as

$$KDP \quad (KH_2PO_4)$$

$$ADP \quad (NH_4H_2PO_4)$$

of symmetry 42m, tetragonal class, uniaxial.

$$\alpha_{41} \quad \text{and} \quad \alpha_{63} \neq 0 \text{ as in quartz.}$$

$$\text{for ADP} \quad n^3 \, \alpha_{41} = 95.10^{-10},$$

one hundred times larger than in quartz.

Still higher values have been found in $LiNbO_3$ and $BaTiO_3$.

THE QUADRATIC ELECTRO-OPTICAL EFFECT

I have already mentioned that in K-Na tartrate salt a quadratic effect is superposed on the linear effect. In centrosymmetrical crystals, where the linear effect is forbidden by symmetry considerations, the quadratic effect alone is present and can become important. The quadratic term can be written

$$\Delta \, (\frac{1}{n^2})_{ij} = \beta_{ijkl} \, E_k \, E_l$$

with introduction of coefficients forming a 4-order tensor. We shall not go into details, but draw your attention to the fact that each change of the six coefficients of the index ellipsoid depends now on two field components E_k and E_l. These components can be the components of the same field vector $\vec{E}$, or they can be produced by two different fields, either in the same direction (E_k, E_k) or in two perpendicular directions (E_k, E_l). For example, one field can be a steady field in time E_o and the other one an alternating field $E_1 \cos \omega t$ which modulates the birefringence. In centrosymmetrical ferroelectric crystals, in the paraelectric phase, like for example KTN $(KTaNbO_3)$, if E_0 = 2000 volts/cm, fields of amplitude E_1 = 10 volts/cm are sufficient to create interest in using these crystalline electro-optical phenomena for light modulation.

THE NATURE AND THE RELAXATION TIME OF THE POCKELS EFFECT

Now we shall say some words about the nature and the relaxation time of the Pockels effect. The discoverers of linear electro-optical birefringence, Roentgen and

Kundt, had recognized its connection to the piezoelectric effect. We remember that the piezoelectric effect discovered by the brothers Curie was the following: a stress applied to a quartz plate produces an electric polarization of the plate and gives rise to electric charges on metal plates covering the faces of the plate.

In the year of that discovery, 1881, Lippmann (11) mentioned that thermodynamics suggested the existence of a reciprocal effect. The application of an electric field to a crystal plate must produce a mechanical stress inside the plate. The theory of this effect was developed by Pockels, and the question arose: is the electric birefringence observed not an indirect effect? The cause may be not the electric field itself but the mechanical stress produced by it. We would have the succession: electric field→mechanical stress→induces a birefringence. This indirect effect could be calculated knowing the piezoelectric coefficients. Pockels showed that the measured birefringence could be explained only partly by it. Another cause remained — as in the Kerr effect of liquids, where electrostriction was not able to explain the optical anisotropy. Two theories were put forward to explain the direct effect.

In 1899 Voigt (12) considered an action of the electric field on the electron orbits inside the atom, these electrons being considered at that time as harmonic oscillators. At about the same time Cotton and Mouton proposed to explain the Kerr effect of liquids by their theory of orientation of electrically and optically anisotropic molecules, a theory which was worked out by P. Langevin in analogy to his theory of paramagnetism. This theory states that, if a field is applied, the total refractivity of the medium remains unchanged:

$$n_x + n_y + n_z = 3n = 2n_o + n_e = C^{te} \qquad \frac{n_e - n}{n_o - n} = -2$$

On the other hand, in Voigt's theory Σn should change. The experiment had to decide.

Another problem had to be considered in this connection: when an electric field is suddenly applied or suddenly suppressed, what is the time of establishment and of vanishing of the birefringence? This time is called the relaxation time τ.

In the case of the indirect piezoelectric effect this time should be of the order of the time of propagation, through the size of the medium, of the elastic deformation, whose velocity is that of sound:

$$v \sim km/sec \approx 10^5 \ cm/sec.$$

This leads to relaxation times of the order of $\tau \sim 10^{-5}$ sec. What is the really observed time? The practical importance of this question is great.

If we want to use the electro-optical effect for light modulation, the upper frequency limit for this use (for the medium to follow the electric impulse) is

$$\omega \sim \frac{1}{\tau} \ .$$

Thus, it is very important to know τ.

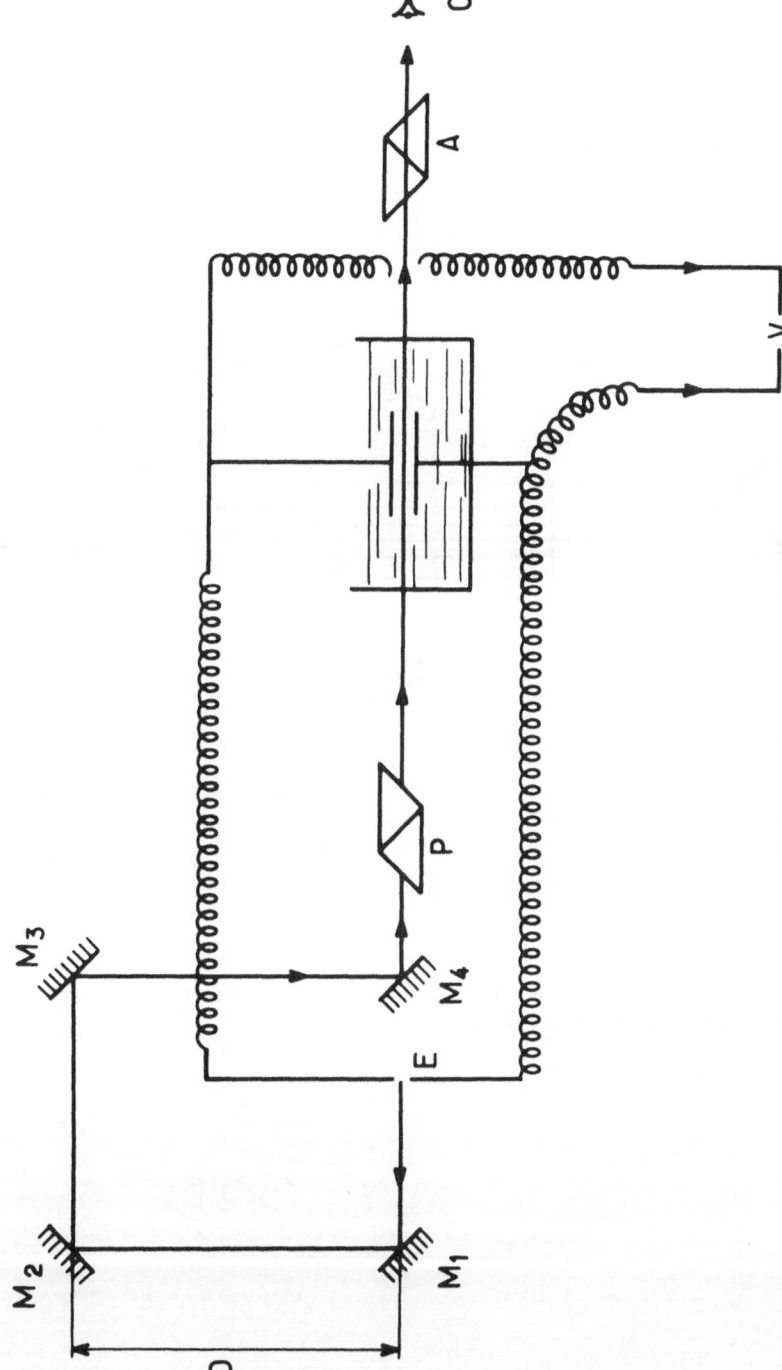

Fig. 1. Scheme of the experiment of Abraham and Lemoine for the measurement of the relaxation time of the Kerr-effect.

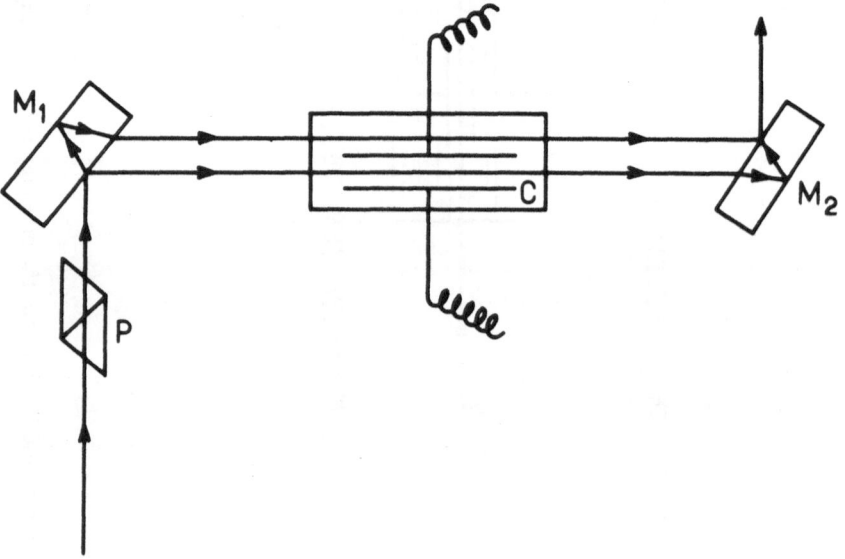

Fig. 2. Experiment of Pauthenier with the Jamin-interferometer: change of the index of refraction by establishing the electric field.

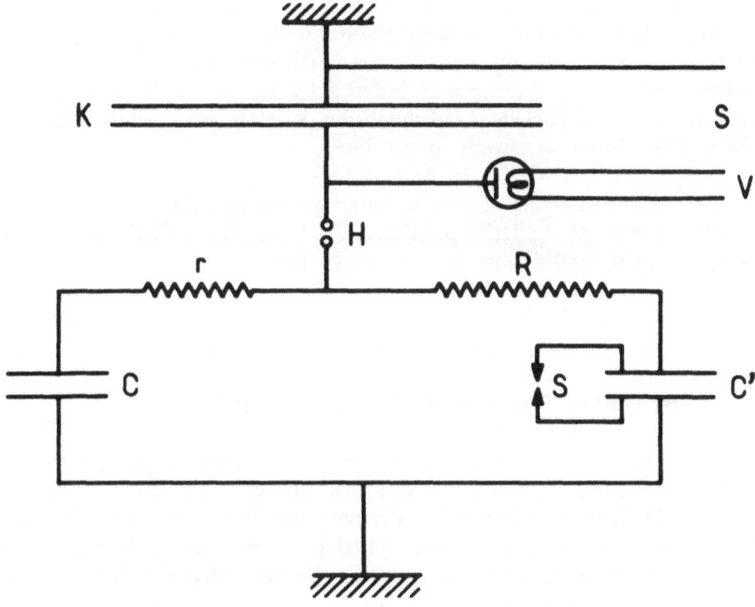

Fig. 3. Electric adjustment in the Pauthenier experiment for avoiding the influence of electrostriction and of Joule-heating of the liquid. The optical measurement must be made immediately after establishing the field. Time delay smaller than 10^{-6} sec.

The first measurement — for the Kerr effect of liquids — was made in the 19th century, which shows that nanosecond techniques were already used at that time. This was the experiment performed by H. Abraham and Lemoine, whose scheme is shown in Figure 1.

It showed that $\tau < 10^{-8}$ sec. It was confirmed later on that τ varies in the range 10^{-10} sec to 10^{-8} sec. We know that in crystals there must be a dispersion effect, a change of the coefficients α and β with applied frequency, since for low frequencies the indirect microscopic stress effect contributes but vanishes for high frequencies.

Later on Pauthenier, inspired by the techniques of Abraham and Lemoine, measured the absolute retardation $n_e - n$ and $n_0 - n$ for liquids by an interference device (Figures 2 and 3) and proved the molecular orientation theory to be correct. The two experiments just mentioned are described in the French treatise on optics by Georges Bruhat (4). The history of optical birefringence induced by an electric field in liquids and in crystals is reviewed in F. Pockels' "Lehrbuch der Kristalloptik", written in German (1). It dates from 1906, but it is always up to date.

The modern aspects of electro-optics in crystals are reviewed in the book "Quantum Electronics" of A. Jariv, published by John Wiley, 1967, in chapter 18, and its application to light modulation, in chapter 19 (3).

Kerr cells of nitrobenzene were used early for light modulation. Crystals were used for the first time by Billings in 1949 (ADP) (13).

STARK EFFECT AND RELATED PHENOMENA

We come now to another electro-optic effect, less important in practice than the effects described, but of great scientific interest. It is the splitting of spectral lines by an electric field. In 1896 Pieter Zeeman had discovered the splitting by magnetic fields called "Zeeman effect". The splitting by electric fields was discovered 19 years later on the atoms of canal rays by the German physicist Johann Stark and at the same time by the Italian physicist Lo Surdo. Both effects are to day explained by quantum mechanics; the theory is based on the quantized energy states of atoms and on the phenomena of space quantization.

Let us review it briefly. An atomic energy state characterized by quantum number J which represents the total angular momentum of the atom, in units $\hbar$, and its magnetic moment in units $g\,\mu_B$, is split in a magnetic field into $2J + 1$ equidistant m-states, numbered from +J to -J. A spectral line is a transition between two energy states, and the splitting of the line is a consequence of the splitting of the two states. The complication of anomalous Zeeman splittings is fully explained by the different values g' and g" (Landé factors) of the two states.

The Zeeman splitting is always symmetrical with respect to the line center (the line position in the absence of a field): $\Delta m = 0$ components are π-polarized ($E \| H_0$) and $\Delta m \pm 1$ components are σ-polarized ($E \perp H_0$) (Figure 4). In small magnetic fields the splitting is always a linear function of H_0. The Stark splitting of a spectral line is also based on the splittings and displacements, produced by an electric field, of the m-substates of

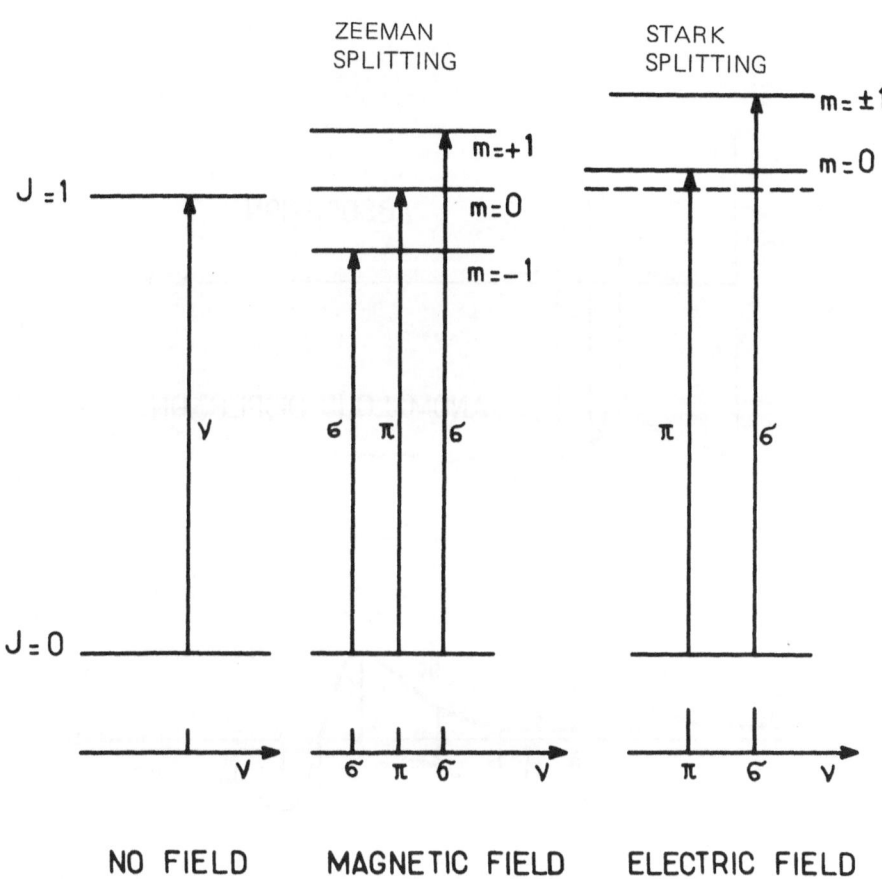

Fig. 4. Zeeman splitting and Stark splitting of a spectral line.

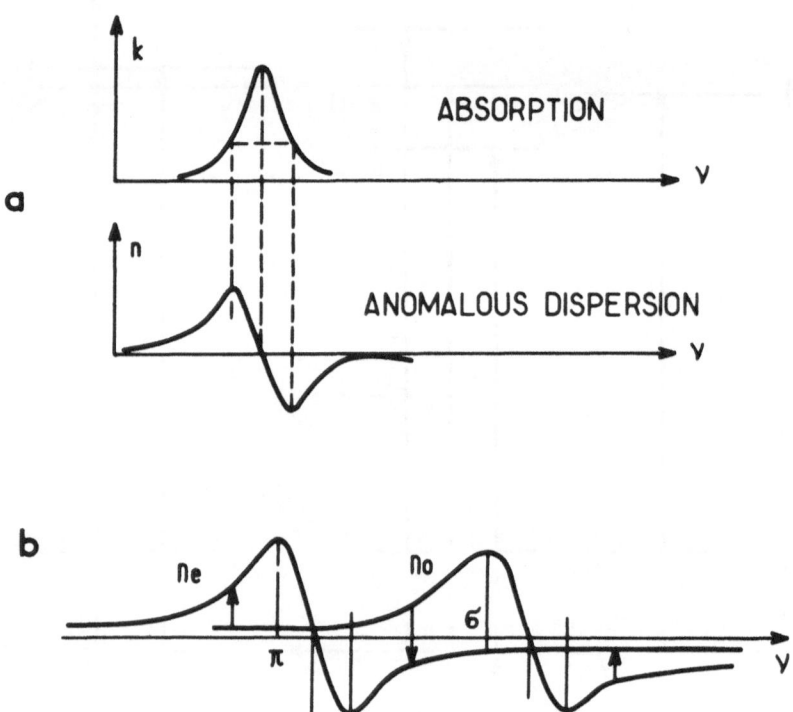

Fig. 5. a) Connection between absorption and anomalous dispersion. b) Relation between Stark-splitting and electric birefringence.

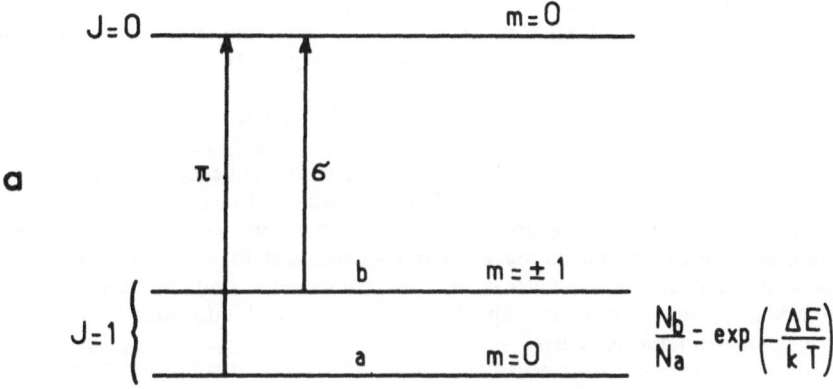

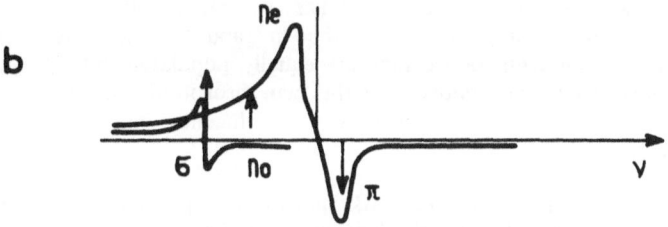

Fig. 6. Birefringence produced by the population change of the Stark levels of the ground-state. a) Stark components. b) Anomalous dispersion lines associated to these components.

the energy states involved.

Two types of splitting must be distinguished: the linear and the quadratic Stark effect. The linear Stark effect is observed on the spectral lines of the hydrogen atom and of hydrogenic atoms. It is a rather exceptional case. In most other cases a quadratic effect, displacements proportional to E^2, is observed. These displacements are not symmetrical with respect to the position of the undisplaced line, and ±m states are displaced at the same rate. Let us take as an example a spectral line representing a transition from a lower J = 0 state to an upper J = 1 state, and compare the splitting of the upper state in a magnetic and in an electric field (Figure 4).

Such split spectral lines can be observed either in emission or in absorption. In general, these lines have a line breadth determined — in gases — by the Doppler effect. We know that the change of refraction index with frequency, dispersion, is governed by the absorption lines of a medium. We have a relation between the absorption curve and the dispersion curve showing anomalous dispersion (Figure 5a). If we have Stark splitting of a line, the Stark components are π- or σ-polarized, and to each component belongs a dispersion curve. In the simple case of two components only we have the following situation, as shown in Figure 5b. To the π-component corresponds the n_e curve; to the σ-component the n_0 curve.

We see that there is a birefringence ($n_e \neq n_0$) of positive sign outside the components, and of negative sign between the two components. This birefringence of the medium, related to the Stark splitting of spectral lines, is just the effect which had been expected by Voigt's classical theory. We may for this reason call it the Voigt effect, but it has to be described by quantum theory. Another effect can be foreseen in atoms and also in paramagnetic ions in crystals at very low temperature, liquid helium temperature.

Suppose that the ground state is a paramagnetic J $\neq$ 0 state, and take a spectral line J = 1 → J = 0 (upper state) which has two Stark components π- and σ-polarized (Figure 6a). The population ratio of the "b" and "a" states is given by Boltzmann's relation. At high temperature they are equally populated, but at low temperature the state above b becomes empty, and the two absorption lines π and σ are of different intensity: $I_\pi > I_\sigma$. Then we will have a birefringence which is positive for $\nu < \nu_0$ and negative for $\nu > \nu_0$ (Figure 6b).

In paramagnetic ions in crystals Stark splittings are produced not only by an applied external electric field, but by the internal crystalline field of electrostatic nature, which can be very high and explains the composite structure of crystalline absorption lines. For example, Ni^{++} in Ni fluosilicate, where the lowest energy state of the ion is a J = 1 state split by the crystalline field into m = 0 and m = ± 1 levels (Figure 7). In this case, at low temperature the birefringence due to population differences will be superposed on the normal birefringence of the crystal. This effect is analogous to the paramagnetic Faraday rotation of light, observed at low temperature.

An analogous "paramagnetic birefringence" can be observed in monatomic vapors whose atoms have been aligned by optical pumping. The birefringence has been observed by Gozzini and used to monitor the degree of alignment. This is, I believe, the early story of electro-optics.

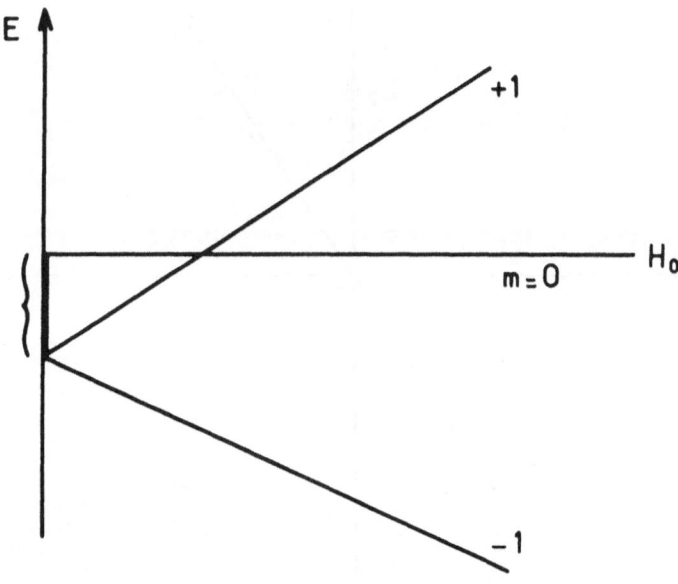

Fig. 7. Ground state of the Ni^{++} ion split by the crystalline field.

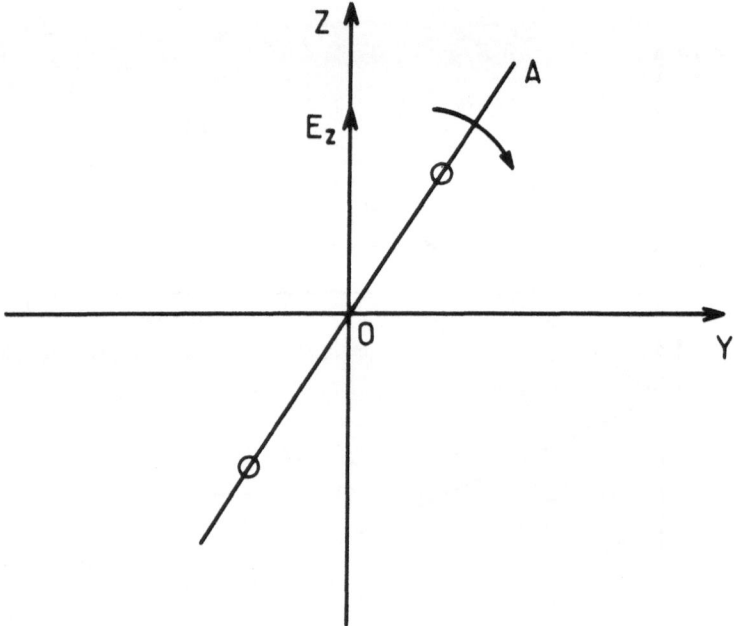

Fig. 8. Rotation of a homopolar diatomic molecule.

It is not my purpose to review its practical applications, especially in the field of light modulation, nor its development in conjunction with laser techniques. May I just, as a last example, cite a case which may be interesting in the field of molecular physics (14).

As is well known diàtomic molecules have quantized vibrational energy states and quantized rotational states whose energy is given by

$$E_J = \frac{h^2}{8 \pi^2 I} \, J \, (J + 1)$$

where h is Planck's constant, J the rotational quantum number, and I the moment of inertia around the axis of rotation, perpendicular to the molecular axis. If the molecule has no center of symmetry, if it is a heteropolar molecule possessing a permanent electric moment, like HCl or ICl, it has vibrational and rotational emission and absorption lines in the infrared or microwave region. If the molecule is homeopolar or centrosymmetrical like H_2, N_2, or O_2 these lines are forbidden. In the classical theory the electromagnetic radiation is related to a change of electric moments, this change being due to the rotation of the permanent moment for the rotational lines.

Consider a molecule like N_2. It has no permanent electric moment, but an electric field $\vec{E}$ induces an induced moment

$$\vec{P} = | \alpha | \vec{E} \quad \text{where} \quad | \alpha |$$

is the polarizability tensor. Such a molecule is anisotropic and has different polarizabilities $\alpha_\parallel$ and $\alpha_\perp$ relative to the molecular axis.

Suppose $\alpha_\parallel \gg \alpha_\perp$ and $\alpha_\perp$ negligible, and suppose we apply a steady electrostatic field E_Z to the nitrogen gas. Consider a molecule whose rotational axis is the X-axis (Figure 8). The rotational circular frequency is ω.

We will have $E_\parallel = E_Z \cos \vartheta$ and $P_\parallel = \alpha_\parallel E_\parallel = \alpha_\parallel E_Z \cos \vartheta$. If we project this on the Z-axis, the Z-component of $\vec{P}$ will be

$$P_Z = \alpha_\parallel \cos^2 \vartheta \, E_Z = \alpha_\parallel \cos^2 \omega t \, E_Z$$

$$\cos^2 \omega t = \frac{1}{2} + \frac{1}{2} \cos 2 \omega t$$

$$P_Z = P^o{}_Z + \frac{1}{2} \alpha_\parallel \, E_Z \cos 2 \omega t$$

This time-dependent moment will radiate. Thus, applying the correspondence principle we expect that in the presence of a strong electric field the rotational lines corresponding to $\Delta J = 2$ will appear in absorption.

Of course, statistically we have to consider all possible orientations in space of the rotational axis. It can be shown that the properties of this absorption, concerning the selection rules and the polarization rules, are the same as for Raman scattering by an incident light vector E_Z. This effect has not been observed, but it is well known that in a gas, at high pressure, these absorption lines appear and can be explained by the action on the molecules of the intermolecular field. This effect could probably be strengthened by adding ions to the gas.

REFERENCES

(1)   F. Pockels, Lehrbuch der Kristalloptik, Teubner, 1906.

(2)   W. Voigt, Kompendium der Kristall - physik, 1896, Chap. V.

(3)   A. Jariv, Quantum Electronics, J. Wiley, 1967, Chapters 18 and 19.

(4)   G. Bruhat, Cours de Physique Genèrale, Optique, Masson 1965, Chap. XXIV, paragraph IIIa, Chap. XXXI.

(5)   Max von Laue, Geschichte der Physik, Univ. Verlag, Bonn, 1947, Chap. 5.

(6)   J. Kerr, Phil. Mag. 1875. 50, 337, 446 a; 1879, 8, 185, 229.

(7)   W. C. Roentgen, 1883, Wied. Ann., 18, 213, 534.

(8)   A. Kundt, (1883), Wied. Ann, 228.

(9)   F. Pockels, (1894), Abb. Goettingen Ges. D. Wiss., 39, 169.

(10)   J. and P. Curie, (1881), C. R. Ac. Sc., 93, 1137, 1881, A; (1882), 95, 914, 1882.

(11)   G. Lippmann, (1881), Ann. Chim. Phys. 24, 145.

(12)   W. Voigt, (1899), Wied. Ann, 69, 297.

(13)   B. Billings, (1949), J. Opt. Soc. Am., 39, 802.

(14)   A. Kastler, (1950), C. R. Ac. Sc., Paris, 230, 1596.

FUNDAMENTALS OF ELECTRO- AND MAGNETO-OPTICS AND NON-LINEAR OPTICS

O. S. Heavens

University of York

Heslington, York (U. K.)

1. BRIEF RÉSUMÉ OF PROPAGATION IN ISOTROPIC AND ANISOTOPIC MEDIA

1.1 Linear, Isotropic, Homogeneous Media

Linearity implies a limitation on the strength of the electric field in the electromagnetic wave. For moderate field strengths, the electric interaction is much stronger than the magnetic. Thus the ratio of magnetic to electric force on an electron subject to a "free space" e-m field is given by

$$\frac{\text{Magnetic force}}{\text{Electric force}} = \frac{e^2 F}{2m^2 \omega^2 c^3 \epsilon_0} \qquad 1.1$$

where F is the magnitude of the Poynting vector. For the visible region of the spectrum this ratio is of the order $10^{-23} F$ where F is in watts/m^2. Thus for any but the highest power fluxes, the magnetic interaction is negligible.

For linear, homogeneous, isotropic media, the propagation of e-m waves implies a power flux given by the Poynting vector $\underline{F}$, where

$$\underline{F} \text{ (watts m}^{-2}) = \underline{E} \text{ (volts m}^{-1}) \text{ x } \underline{H} \text{ (amps m}^{-1}) \qquad 1.2$$

For moderate field strengths, insufficient to introduce non-linear behaviour in the medium of propagation,

$$\underline{D} = \epsilon \underline{E} = \epsilon_r \epsilon_0 \underline{E}$$
$$\underline{B} = \mu \underline{H} = \mu_r \mu_0 \underline{H} \qquad 1.3$$

19

ϵ and μ are scalars, $\underline{D} \parallel \underline{E}$, $\underline{B} \parallel \underline{H}$ and for optical frequencies, μ_r may be taken as unity. The velocity of propagation (phase velocity) is $(\mu \epsilon)^{-\frac{1}{2}}$ which for free space becomes $(\mu_0 \epsilon_0)^{-\frac{1}{2}} = 3.0 \times 10^8 \text{ m s}^{-1}$. $|\underline{E}| / |\underline{H}| = 377$ ohms.
The refractive index n is given by

$$n = \frac{(\mu \epsilon)^{\frac{1}{2}}}{(\mu_0 \epsilon_0)^{\frac{1}{2}}} \simeq (\epsilon / \epsilon_0)^{\frac{1}{2}} = \epsilon_r^{\frac{1}{2}}, \quad \text{where the}$$

permittivity ϵ_r refers only to the electronic contribution to the total permittivity of the medium. For ionic and polar materials, the low-frequency permittivity may be very much higher than n^2.

1.2   Propagation in Anisotropic Media

At optical frequencies, $\mu_r = 1$: anisotropy results from ϵ. $\underline{B} \parallel \underline{H}$ but $\underline{D}$ is not generally parallel to $\underline{E}$. ϵ is a tensor. The energy density W_E associated with a field of magnitude $|\underline{E}|$ depends on the direction of $\underline{E}$, through the relation

$$W_E = \frac{1}{2} \left\{ \epsilon_{XX} E_X^2 + \epsilon_{YY} E_Y^2 + \epsilon_{ZZ} E_Z^2 + 2 \epsilon_{XY} E_X E_Y \right.$$

$$\left. + 2 \epsilon_{YZ} E_Y E_Z + 2 \epsilon_{ZX} E_Z E_X \right\} \qquad 1.4$$

Choose new axes (x,y,z) parallel to the principal axes of the ellipsoid $W_E = $ const., so that

$$W_E = \frac{1}{2} \left\{ \epsilon_x E_x^2 + \epsilon_y E_y^2 + \epsilon_z E_z^2 \right\} \qquad 1.5$$

ϵ_x, ϵ_y and ϵ_z are termed principal permittivities. Corresponding refractive indices are $n_x = \epsilon_x^{\frac{1}{2}}$, $n_y = \epsilon_y^{\frac{1}{2}}$ and $n_z = \epsilon_z^{\frac{1}{2}}$ and are functions of frequency.

Wave Surfaces

Plane-wave solutions exist for the anisotropic medium, in which the normal to the wavefront is in the direction $\underline{D} \times \underline{H}$. The direction of energy flow (Poynting vector-ray) is $\underline{E} \times \underline{H}$. Medium characterised by three indices n_x, n_y, n_z and corresponding wave velocities v_x, v_y, v_z where $v_i = c/n_i$, i = x,y,z. The wave velocity b in a direction (S_x, S_y, S_z) is given by

$$\sum_{i = x,y,z} \frac{S_i^2}{b^2 - v_i^2} = 0 \qquad 1.6$$

which is quadratic in b^2 and gives in general two velocities for a given direction, and hence two wavefronts. If $v_x = v_y$, there is one spherical and one spheroidal wave surface, with a common v_z-axis, viz:

$$r^2 = v_z^2 \qquad \qquad \text{(Sphere)}$$

$$\text{and} \quad \frac{x^2}{v_z^2} + \frac{y^2}{v_x^2} + \frac{z^2}{v_z^2} = 1 \qquad \text{(Spheroid)} \qquad \qquad 1.7$$

(<u>uniaxial</u> media).

In general, the wave surface is given by the quartic equation

$$r^2 (v_x^2 x^2 + v_y^2 y^2 + v_z^2 z^2) - v_x^2(v_y^2 + v_z^2) x^2 - v_y^2(v_z^2 +$$

$$v_x^2) y^2 - v_z^2(v_x^2 + v_y^2) z^2 + v_x^2 v_y^2 v_z^2 = 0 \qquad \qquad 1.8$$

In each of the three coordinate planes there is one circular an one elliptic intersection with the wave surface (biaxial media) (Fig. 1). The case of gyrotropic media is dealt with in section 3.

The wave surface(s) described by equations (1.7) and (1.8) may be regarded as the envelopes of sets of plane waves in all directions, starting at a point 0 in the medium at t = 0. The point P at which the plane wave in a given direction touches the wave surface defines the direction OP of the ray corresponding to the given plane wave. Thus the surfaces (1.7), (1.8) are in this sense ray surfaces. In all directions other than those of the coordinate axes used to define W_E in equation (1.5), the directions of ray and wave-normal do not coincide.

Conventionally, the directions of the coordinate axes are taken so that n_y lies between n_x and n_z. With this choice, equation (1.6) shows that the two values of phase velocity coincide for the direction

$$S_x = \left(\frac{v_x^2 - v_y^2}{v_x^2 - v_z^2} \right)^{\frac{1}{2}}, \quad S_y = 0, \quad S_z = \left(\frac{v_y^2 - v_z^2}{v_x^2 - v_z^2} \right)^{\frac{1}{2}} \qquad 1.9$$

The plane wave in this direction touches the quartic surface (1.8) in a circle of contact, providing a simple and elegant explanation of the phenomenon of internal conical refraction.

Rays

The velocity B of a ray in the direction (σ_x, σ_y, σ_z) is given by

$$\sum_{i = x,y,z} \frac{\sigma_i^2 v_i^2}{B^2 - v_i^2} \qquad \qquad 1.10$$

Coincident roots of this equation serve to define the axes of single ray velocity, which occur in the directions

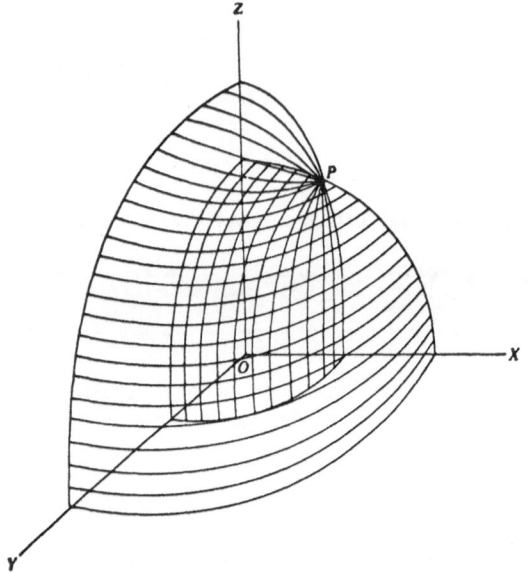

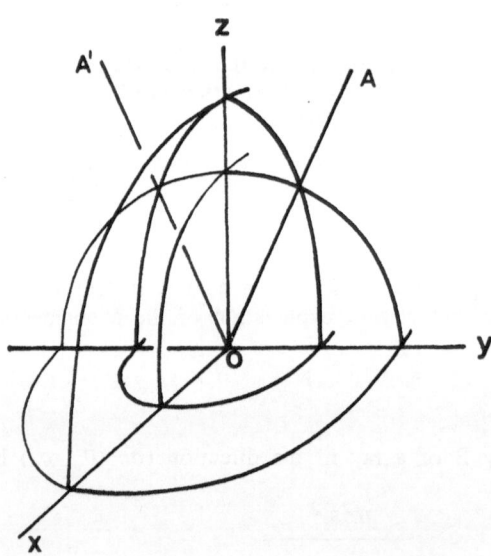

Fig. 1. Wave surface for biaxial media.

$$\sigma_x = \frac{v_z}{v_x} \left(\frac{v_x^2 - v_y^2}{v_x^2 - v_z^2} \right)^{\frac{1}{2}}, \ \sigma_y = 0, \ \sigma_z = \frac{v_x}{v_z} \left(\frac{v_y^2 - v_z^2}{v_x^2 - v_z^2} \right)^{\frac{1}{2}}$$

1.11

It is seen that the complete optical behaviour of the linear, anisotropic, homogeneous medium may be deduced once the principal refractive indices, and therefore the corresponding velocities (v_x, v_y, v_z) are known. It should be emphasised that these quantities are frequency-dependent and that the directions of the optic axes of a biaxial material generally depend on both frequency and temperature. This phenomenon is of special importance in phase-matching in the generation of harmonics, as described in section 6.3

Index Ellipsoid

The principal refractive indices n_x, n_y, n_z serve to define an index ellipsoid, or indicatrix. This may be used to determine the phase velocities for a given direction of wave-normal inside the medium. A plane through the centre of the ellipsoid, and parallel to the wave-front of interest, intersects the index ellipsoid in an ellipse. If the major and minor axes of the elliptic section are n′ and n″, then the phase velocities of the phase velocities of the two waves in the direction of interest are c/n′ and c/n″.

2.   CRYSTAL CLASSES EXHIBITING DOUBLE REFRACTION

Crystals belonging to the cubic system are optically isotropic (although they may be rendered anisotropic by the application of an electric field - see section 4). Other crystal classes are anisotropic according to the following scheme:

Uniaxial	Biaxial
Trigonal	Orthorhombic
Tetragonal	Monoclinic
Hexagonal	Triclinic

In uniaxial crystals, the optic axis lies along the crystal axis of highest symmetry. The directions of the optic axes in biaxial materials obey the following rules:

Orthorhombic: semi-axes of the index ellipsoid are parallel to the crystal axes (so that the optic axes lie in one of the coordinate planes).

Monoclinic: optic axes either lie in the plane perpendicular to the 2-fold axis or optic axes and 2-fold axis lie in a plane.

Triclinic: no constraints.

A feature of the materials described above is that there exist directions along which a beam of plane polarised radiation propagates without change of the state of polarisation. (The amplitude of the beam may decrease, when absorption occurs: this may be taken into account by allowing the refractive index to assume complex values. The state of polarisation however is not changed in this case). There is a class of materials for which this does not apply. The phenomenon of optical activity has generally been described as one in which the plane of polarization of plane polarized light is rotated about the direction of propagation as the radiation travels through the medium. In fact this is an oversimplified description. Light which is initially plane polarised generally becomes elliptically polarised, with the plane of the major axis rotated with respect to that of the original plane of polarisation. Such media are described as gyrotropic and their properties are investigated in the following section.

3.   GYROTROPIC MEDIA

3.1   Propagation

In contrast to the media discussed in sections 1 and 2, gyrotropic media may be described phenomenologically by skew-symmetric permeability and permittivity tensors. Although the permeability at optical frequencies can often be taken as unity, so that gyrotropy results only from the permittivity tensor, this is not invariably the case. The behaviour of the ferromagnetic metals Fe, Ni and Co requires the assumption that both ϵ and μ are skew-symmetric tensors. It is likely however that for materials whose electro-optic behaviour is of interest, only the permittivity will take the tensor form. This will be assumed in the development which follows: We assume a permittivity tensor of the form

$$\left[\epsilon\right] = \begin{bmatrix} \epsilon_q & -i\,\epsilon_q q & 0 \\ i\epsilon_q q & \epsilon_q & 0 \\ 0 & 0 & \epsilon \end{bmatrix} \qquad\qquad 3.1$$

where the gyroelectric constant q may be complex, as may also the diagonal elements ϵ_q and ϵ.

Maxwell's equations for a gyroelectric medium are

$$\left. \begin{aligned} \nabla \times \underline{H} &= \left[\epsilon\right] \dot{\underline{E}} \\ \nabla \times \underline{E} &= -\mu\, \dot{\underline{H}} \end{aligned} \right\} \qquad\qquad 3.2$$

For the plane wave solution, with

$$\underline{E}\,(\underline{r},\, t) = \underline{E}\, \exp\left[-i\,\omega\,(t - \frac{n}{c}\,\underline{r}\cdot\underline{s})\right]$$

$$\underline{H}\,(\underline{r},\, t) = \underline{H}\, \exp\left[-i\,\omega\,(t - \frac{n}{c}\,\underline{r}\cdot\underline{s})\right] \qquad\qquad 3.3$$

where $\underline{s}$ is a unit vector, direction cosines α, β, γ, we have

$$-\frac{n}{c} \ (\underline{s} \times \underline{\underline{H}}) \ = \ [\,\epsilon\,] \ \underline{\underline{E}}$$

and
$$\frac{n}{c} \ (\underline{s} \times \underline{\underline{E}}) \ = \ \mu\underline{\underline{H}}$$

3.4

which leads to

$$n_{\pm}^2 \ [\underline{\underline{E}} - \underline{s} \ (\underline{s} \cdot \underline{\underline{E}})] \quad = \quad c^2 \ \mu \ [\,\epsilon\,] \ \underline{\underline{E}}$$

3.5

where the symbol $n_{\pm}$ indicates that, as shown below, there are two possible indices of refraction corresponding to a given direction of propagation. Equation (3.5) may be expanded as

$$n_{\pm}^2 \ [E_x - \alpha(\alpha E_x + \beta E_y + \gamma E_z)] \ = \ \eta^2(E_x - iqE_y)$$

$$n_{\pm}^2 \ [E_y - \beta(\alpha E_x + \beta E_y + \gamma E_z)] \ = \ \eta^2(igE_x + E_y)$$

$$n_{\pm}^2 \ [E_z - \gamma(\alpha E_x + \beta E_y + \gamma E_z)] \ = \ n^2 E_z$$

3.6

where $n = c \ (\mu \ \epsilon_q)^{\frac{1}{2}}$ and $\eta = c \ (\mu \ \epsilon)_q^{\frac{1}{2}}$

Equations (3.6) are consistent only if the determinant vanishes, to yield

$$n_{\pm}^4 \ [\eta^2 \ (1 - \gamma^2) + n^2\gamma^2] - n_{\pm}^2 \ [\eta^2 n^2(1 + \gamma^2) + \eta^4(1 - \gamma^2) \ (1 - q^2)]$$

$$+ \ \eta^4 n^2 (1 - q^2) + 0$$

3.7

The general solution of (3.7) for $n_{\pm}^2$ is straightforward, if somewhat clumsy. Since one will generally be concerned with the incidence of radiation on a gyroelectric medium (rather than solely by propagation in the medium) it is convenient to consider three special cases of orientation of the gyroelectric axis with respect to the surface and plane of incidence. These are (Fig.2):

(i)   Transverse case: gyroelectric axis (0z) perpendicular to plane of incidence so that $\gamma = 0$;

(ii)   Longitudinal case: gyroelectric axis (0z) parallel to plane of incidence so that $\alpha = 0$ (0z) $\perp$ plane of incidence).

(iii)   Polar case: gyroelectric axis (0z) perpendicular to plane of incidence (again, choose Ox $\perp$ plane of incidence : $\alpha = 0$).

For the transverse case ($\gamma = 0$), equation (3.7) reduces to

$$n_+ = n_- = \ \eta(1 - \frac{q^2}{2})^{\frac{1}{2}}$$

3.8

The general expressions for the longitudinal and polar cases are somewhat cumbersome. However, for the optical region of the spectrum, and for materials with an initially

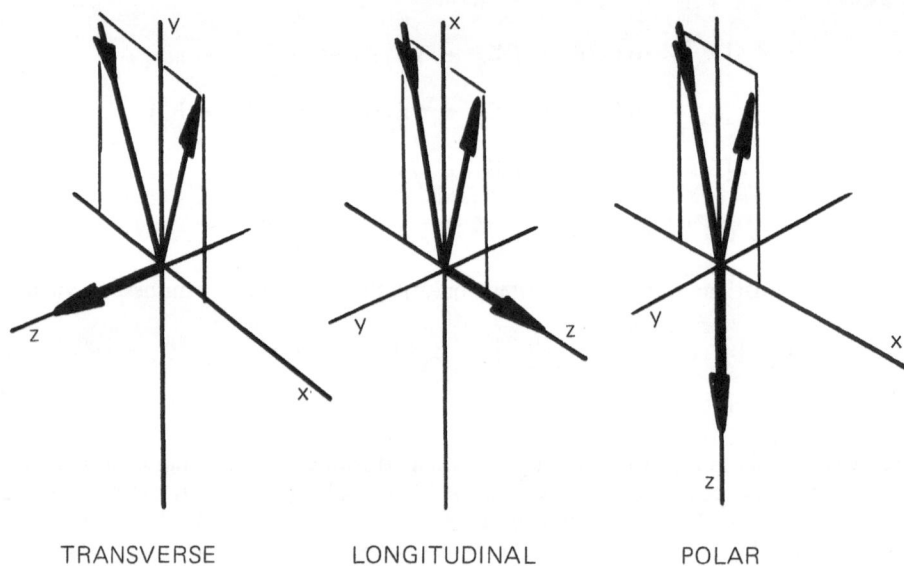

Fig. 2. Gyrotropic media: transverse, longitudinal and polar cases.

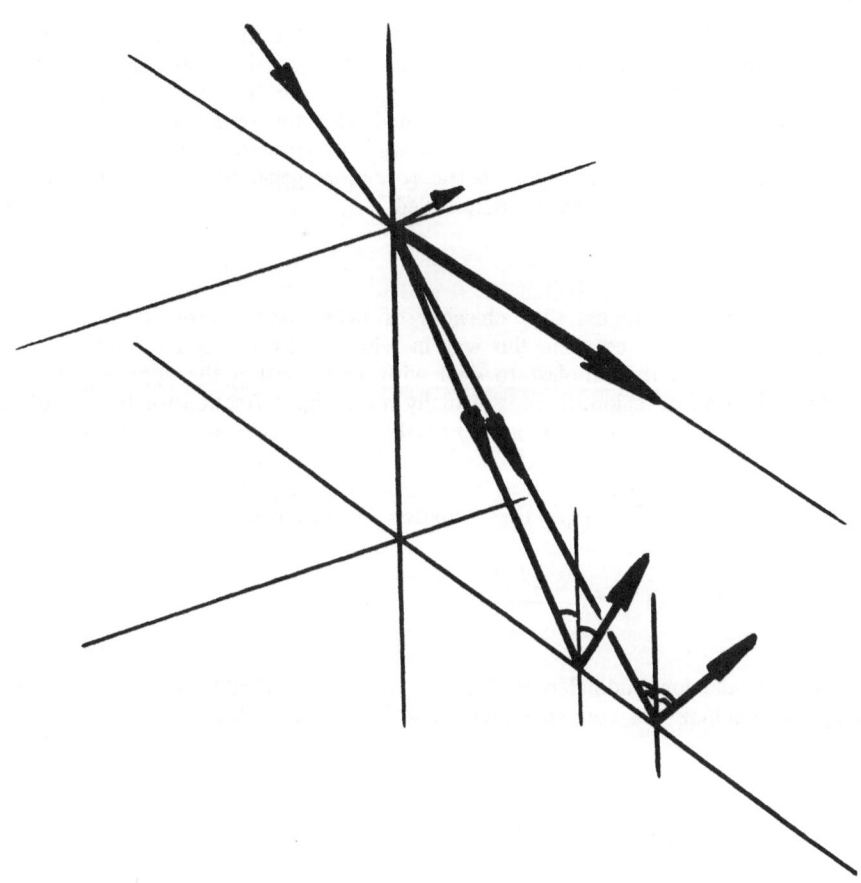

Fig. 3. Double refraction in longitudinal case.

cubic symmetry, the difference between ϵ_p and ϵ is negligible so that where appropriate, the approximation $n \simeq \eta$ may be used.

In this case, we obtain

$$n_\pm^2 = n^2 [1 - \tfrac{1}{2}\beta^2 q^2 \pm q \sqrt{\gamma^2 + \tfrac{1}{4}\beta^2 q^2}] \qquad\qquad 3.9$$

The tensor (3.1) which forms the basis of the above analysis applies to an initially isotropic material, of which the anisotropy results from the presence of gyroelectric effects. It could also apply, with $\epsilon_p \neq \epsilon$, to the case of a gyroelectric material in which the gyroelectric axis coincided with the optic axis of the material, as could arise in crystals of tetragonal or hexagonal symmetry. The general case of a biaxial material which is also gyroelectric (or, more generally, gyroelectric and gyromagnetic) would be somewhat complicated.

3.2   Reflection and Refraction

In order to discuss the behaviour of light passing through gyroelectric elements, it is necessary to examine the way in which reflection and refraction occur. The transverse case is the simplest to deal with since, within the approximations appropriate to the optical region, these is virtually no double refraction for this condition. Thus for the case $\epsilon_q = = \epsilon$, equation (3.8) shows that $n_+ = n_- = n (1 - \frac{q}{2})^{\tfrac{1}{2}}$

The polar and longitudinal cases are more involved. Before discussing these, we note that in general $q \ll 1$, so that equation (3.9) may be simplified to read

$$n_\pm^2 = n^2 (1 \pm \gamma q - \frac{\beta^2 q^2}{2}) \qquad\qquad 3.10$$

If we consider the longitudinal case (Fig. 3), we see that double refraction occurs on entering the medium, the corresponding refractive indices being

$$n_+ = n (1 + \gamma q - \frac{\beta^2 q^2}{2})^{\tfrac{1}{2}}$$

$$\qquad\qquad 3.11$$

$$\text{and} \quad n_- = n (1 - \gamma q - \frac{\beta^2 q^2}{2})^{\tfrac{1}{2}}$$

On reflection at the second boundary, the direction of the reflected beam must be such as to satisfy the usual boundary conditions. Since the sign of γ does not change on reflection, then the refractive indices for the reflected beams are the same as for the incident beams and each component is reflected normally.

For the polar case (Fig. 4), we note the change of sign in γ on reflection in the medium. Thus if a beam with wavenormal in direction $(0, \beta, \gamma)$, for which the refractive index of the medium is

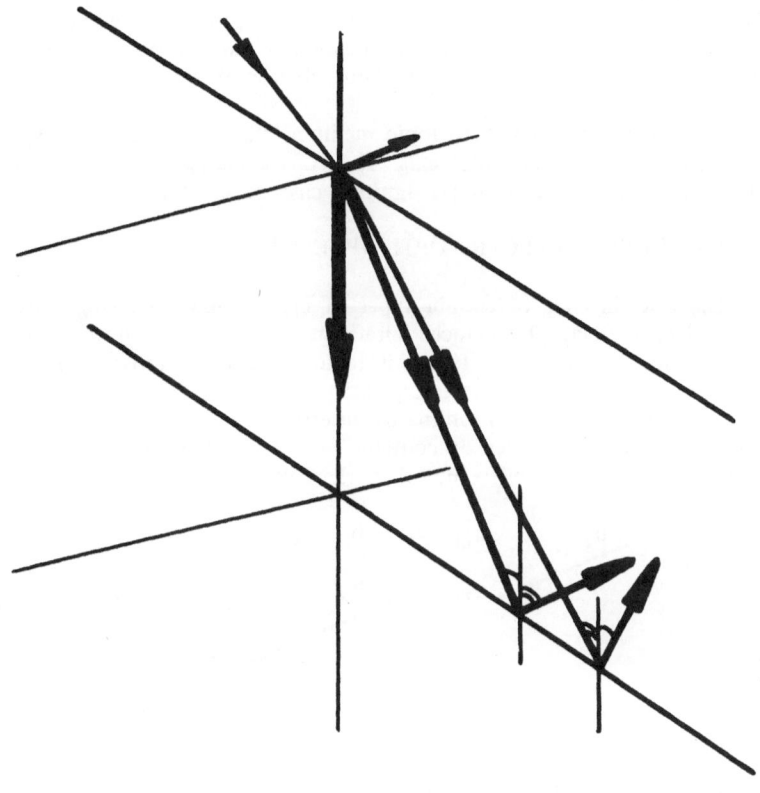

Fig. 4. Double refraction in polar case.

$n^I_{\pm} = n \left(1 \pm \gamma q - \dfrac{\beta^2 q^2}{2}\right)^{\frac{1}{2}}$, is reflected in the medium, the refractive indices for the reflected beams become $n^R_{\pm} = n \left(1 \mp \gamma q - \dfrac{\beta^2 q^2}{2}\right)^{\frac{1}{2}}$. The angles of reflection are thus not equal to the angles of incidence, and the beams couple in the way indicated in Fig. 4.

In conclusion, it may be noted that for a uniaxial, non-gyrotropic medium plane-polarised light in the direction of the optic axis propagates without change of the state of polarisation. For a gyrotropic medium, circularly polarised light propagates in the direction of the gyrotropic axis without change of state of polarisation.

For the uniaxial non-gyrotropic medium, the above comment applies to directions perpendicular to the optic axis. The corresponding result for a gyrotropic medium does not apply, except in the limiting case of $\epsilon_q = \epsilon$.

4.   THE LINEAR ELECTRO-OPTIC (POCKELS) EFFECT

The birefringence of certain types of crystal may be changed by the application of an electric field. The induced birefringence is directly proportional to the applied field $\underline{E}$. Since a second-rank tensor is required to characterise ordinary birefringence and since applied field is a vector quantity, this (Pockels) effect requires a third-rank tensor for specification. For ordinarily birefringent materials (i.e. excluding optical activity) the second-rank tensor contains only six independent components, since $a_{ij} = a_{ji}$. Instead of a double suffix notation, the following notation is employed:

$$
\begin{array}{llll}
a_{11} \equiv b_1 & a_{12} \equiv b_6 & \leftarrow\ a_{13} \equiv b_5 \\[2mm]
\bullet & \quad\ \ a_{22} \equiv b_2 & a_{23} \equiv b_4 \\[2mm]
\bullet & \quad\ \ \ \bullet & a_{33} \equiv b_3
\end{array}
$$

where the arrows serve to indicate the order in which the terms b_i are defined. Thus the equation of the index ellipsoid may be written

$$b_1 X^2 + b_2 Y^2 + b_2 Z^2 + 2b_4 YZ + 2b_5 ZX + 2b_6 XY = 1 \qquad\qquad 4.1$$

When transformed to axes parallel to the principal axes of the ellipsoid, giving

$$B_1 x^2 + B_2 y^2 + B_3 z^2 = 1 \qquad\qquad 4.2$$

the values of B are related to the principal refractive indices by

$$B_i = 1/n_i^2 \qquad (i = 1, 2, 3,).$$

Application of an electric field $\underline{E}$ causes changes in b_i, where

$$
\begin{pmatrix} \Delta b_1 \\ \Delta b_2 \\ \Delta b_3 \\ \Delta b_4 \\ \Delta b_5 \\ \Delta b_6 \end{pmatrix}
=
\begin{pmatrix} r_{11} & r_{12} & r_{13} \\ r_{21} & r_{22} & r_{23} \\ r_{31} & r_{32} & r_{33} \\ r_{41} & r_{42} & r_{43} \\ r_{51} & r_{52} & r_{53} \\ r_{61} & r_{62} & r_{63} \end{pmatrix}
\begin{pmatrix} E_1 \\ E_2 \\ E_3 \end{pmatrix}
\qquad 4.3
$$

where E_1, E_2, E_3 are the components of $\underline{E}$. The effects of crystal symmetry are such as to reduce the number of independent components of r_{ij}. Thus for ammonium dihydrogen phosphate (ADP), which belongs to class $\overline{4}2m$, the only non-vanishing elements r_{ij} are r_{41} r_{25} ($= r_{41}$) and r_{63}. ADP is tetragonal. If the applied electric field is in the direction of the four-fold axis (0z), the principal refractive indices for the resulting biaxial crystal are

$$
\left.\begin{aligned}
n_x &= n_o \left(1 + \tfrac{1}{2}n_o^2 \, r_{63} \, E_z\right) \\
n_y &= n_o \left(1 - \tfrac{1}{2}n_o^2 \, r_{63} \, E_z\right) \\
n_z &= n_o
\end{aligned}\right\} \qquad 4.4
$$

Typical values of r_{63} and r_{41} for the dihydrogen phosphates are given in Table 4.1, from Harvey's "Coherent Light". For light traversing a crystal of thickness d in the direction 0z, the path difference introduced between the x-and y-components of the light field is

$$
\Delta = (n_x - n_y)\, d = n_o^3 r_{63} \, E_z \, d \qquad 4.5
$$

If a voltage V is applied between the ends of the crystal, $E_z = V/d$, so that $\Delta = n_o^3 r_{63} V_z$. The voltage $V_{\lambda/2}$ needed to make $\Delta = \lambda/2$ is given by

$$
V_{\lambda/2} = \frac{\lambda}{2n_o^3 \, r_{63}} \qquad 4.6
$$

and is an important quantity in relation to the use of electro-optical materials as modulators. Since the value of n_o for the dihydrogen phosphates does not vary much over the range of transparency of the crystals, equation (4.6) enables the half-wave voltage to be formed for any wavelength if r_{63} is known.

Certain ferroelectric crystals, such as $BaTiO_3$, $LiNbO_3$, $LiTaO_3$ and mixed oxides such as $KTa_{0.65}Nb_{0.35}O_3$, have assumed considerable importance as (linear) electro-optical materials. Their structures are generally slightly deformed perovskite, the deformation from strictly cubic symmetry resulting from their ferroelectric character. (It is indeed only from such deformation that linear electro-optic behaviour is mani-

Table 4.1 TYPICAL VALUES OF r_{63} AND r_{41} FOR THE DIHYDROGEN PHOSPHATES

Crystal	Temp. (^{o}K)	n_o at 550THz	r_{63} (pm/V)	r_{41} (pm/V)
ADP	148	1.48	−8.5	+24.5
KDP	123	1.47	−10.5	+8.6
KD*P	222	1.47	−24.0	+8.8

Table 4.2 VALUES OF r_{13} AND r_{33} FOR SOME CRYSTALS

Crystal	n_1	n_2	r_{13} (pm/V)	r_{33} (pm/V)
$BaTiO_3$	2.420	2.350	8.0	28
$LiNbO_3$	2.286	2.200	8.6	31
$KTa_{0.65}Nb_{0.35}O_3$	2.290	−	−	300

fested). The only non-vanishing coefficients in the r_{ij} matrix (equation (4.3)) are r_{42}, r_{22}, r_{13} and r_{33}. The crystal is uniaxial and remains so, but with a change of birefringence, when an electric field is applied along the 3-fold axis. For light polarised parallel to, or perpendicular to, this axis, the induced birefringence for a field E is given by:

$$\Delta n^{'} \; - \; \Delta n^{''} \; = \; \tfrac{1}{2}\,(n_1^{\;3} \; r_{33} \; - \; n_2^{\;3} \; r_{13}) \hspace{3cm} 4.7$$

where n_1, n_2 are the refractive indices for the ordinary and extraordinary waves in the direction perpendicular to the optic axis for the field-free case. Values of r_{13}, r_{33}, for one or two materials of interest are shown in Table 4.2

5.   THE QUADRATIC ELECTRO-OPTIC (KERR) EFFECT

5.1   Liquids

Arises from partial alignment of anisotropic molecules by applied electric field. Initially isotropic medium becomes uniaxial with optic axis along field direction. Birefringence is proportional to the square of the fieldstrength. Since the extent of reorientation is governed by thermal relaxation effects, the Kerr effect is temperature-dependent.

For light propagating in direction $\perp$ field direction, the optical path difference for a path ℓ is given by

$$(n_e \; - \; n_o) \; \ell \; = \; B\lambda_o \, E^2 \hspace{3cm} 5.1$$

which serves to define the Kerr constant B. The induced birefringence may be either positive ($n_e > n_o$) or negative. Typical values for a temperature of 20°C and wavelength 3.7 μm are shown in Table 5.1.

5.2   Crystals

Effects in anisotropic crystals very complicated. Changes in the constants of the index ellipsoid follow the relation

$$\Delta(1/n_r^{\;2})\,ij \;\; = \;\; g_{ijkl}P_kP_l \hspace{3cm} 5.2$$

where P_k, P_l are the polarisation components and g_{ijkl} is the electro-optic tensor.

For cubic materials (0_h) the tensor possesses only the following non-zero components   5.3

$$g_{11}, \quad g_{22} \;\; = \;\; g_{33} \;\; = \;\; g_{11},$$

$$g_{12}, \quad g_{13} \;\; = \;\; g_{21} \;\; = \;\; g_{23} \;\; = \;\; g_{31} \;\; = \;\; g_{32} \;\; = \;\; g_{12}$$

$$g_{44}, \quad g_{55} \;\; = \;\; g_{66} \;\; = \;\; g_{44}$$

Table 5.1 TYPICAL VALUES OF n_o AND B FOR SOME LIQUIDS

Liquid	n_o	B $(nm/V)^2$
Nitrobenzene	1.55	44000
Acetone	1.36	1800
Water	1.33	520
Chloroform	1.45	-370
Benzene	1.51	70

If an electric field is applied in the 0z direction of a cubic crystal, parallel to a cubic crystal, parallel to a cube edge, the index ellipsoid takes the form

$$(\frac{1}{n_o{}^2} + g_{12} P_z{}^2) (x^2 + y^2) + (\frac{1}{n_o{}^2} + g_{11} P_z{}^2)z^2 = 1 \qquad 5.4$$

Thus the optical path difference introduced in the components of a beam travelling perpendicular to the field direction is

$$\ell \Delta n = \frac{n_o{}^3 \ell}{2} (g_{11} - g_{12}) (\epsilon - 1)^2 E_z{}^2 \qquad 5.5$$

where ϵ is the permittivity of the crystal. For $BaTiO_3$, the value of $(g_{11} - g_{12})$ is $0.13 m^4/c^2$ and n_o is 2.4 and $\epsilon \sim 5000$. From the above figures it appears that a potential difference of 1.3 kV across a $BaTiO_3$ crystal 1 mm thick should suffice to produce a path difference of a half wavelength for the visible spectrum.

6.   NON–LINEAR OPTICAL PHENOMENA

6.1   Order of Magnitude of Field Required

In the previous sections, we have dealt with optical anisotropy, either naturally-occurring, or induced by the application of electric or magnetic fields. (Since the symposium is concerned with electro-optics. The Cotton-Mouton, Faraday and Kerr magneto-optical effects have been omitted). In the above treatments it has been assumed that the electromagnetic radiation is linear, an assumption which is justified for intensities which are sufficiently low.

We can get an idea of the flux density in a radiation field at which non-linear effects may be expected. Consider, e.g., the case of ADP for which, writing the polarisation in the form

$$P = \epsilon_o \epsilon_r E (1 + \chi^{(1)} E + \chi^{(2)} E^2 + \ldots \ldots) \qquad 6.1$$

the value of $\chi^{(1)}$ is approximately 5.9×10^{-13} in m.k.s. units. Thus the field strength E at which the second term becomes 10^{-6} is $E = 1.7 \times 10^6$ volts/m. In ADP, this would correspond to a flux density, for a uniform field, of 1.7×10^{10} watts/m². Since the arrival of the laser, power densities vastly in excess of this value have been easily attainable, and a wealth of non-linear effects demonstrated and utilised.

6.2   Effect of Anharmonic Response

Classical dispersion theory, first order, analyses the one-dimensional problem of an electric field E applied to a charge e, with damping, by solving the equation

$$\ddot{x} + \gamma \dot{x} + \omega_o^2 x = \frac{e}{m}(E + \frac{P}{3\epsilon_o}) \qquad 6.2$$

where the usual Lorenz internal field correction is assumed to apply. For an applied field of the form $E = E_o \exp(i\omega t)$, where E_o may be complex, the resulting polarisation is given by

$$P = \frac{Ne^2}{m} \frac{1}{[(\omega_o^1)^2 - \omega^2] + i\gamma\omega} E_o e^{i\omega t} \qquad 6.3$$

where $(\omega_o^1)^2 = \omega_o^2 - Ne^2/3m\epsilon_o$.

In order to examine the effect of non-linearities, we assume that the motion of the electron obeys the equation

$$\ddot{x} + \gamma \dot{x} + \omega_o^2 x - \zeta x^2 = \frac{e}{m}(E + \frac{P}{3\epsilon_o}) \qquad 6.4$$

where the anharmonic term ζx^2 is small compared with the other terms on the l.h.s. We set $x = \alpha_1 E + \alpha_2 E^2$, so that the fundamental and second-harmonic polarisabilities $P_{(1)}$ and $P_{(2)}$ are given by

$$P_{(1)} = Ne\alpha_1 E \qquad P_{(2)} = Ne\alpha_2 E^2 \qquad 6.5$$

On solving equation (6.4) and collecting terms of similary order, we obtain

$$\ddot{P}_{(1)} + \gamma \dot{P}_{(1)} + (\omega_o^1)^2 P_{(1)} = \frac{Ne^2 E}{m} \qquad 6.6.$$

$$\ddot{P}_{(2)} + \gamma \dot{P}_{(2)} + (\omega_o^1)^2 P_{(2)} = \frac{\zeta}{Ne} P_{(1)} \qquad 6.7$$

The effect of non-linearities is to cause mixing of the frequencies of any e-m fields present. Thus we consider the general case where

$$E = \sum_n E_n(\omega_n) \exp(i\omega_n t) \qquad 6.8$$

Solution of equation (6.6) for this value of E yields

$$E_{(1)} = \frac{Ne^2}{m} \sum_n \frac{E_n(\omega_n) \exp(i\omega_n t)}{(\omega_o^1)^2 + i\gamma\omega_n - \omega_n^2} \qquad 6.9$$

To obtain $P_{(2)}$, the value of $P_{(1)}$ above is substituted into equation (6.7) to give

$$P_{(2)} = \frac{\zeta Ne^3}{m^2} \sum_n \sum_m \frac{E_n(\omega_n) E_m(\omega_m) \exp i(\omega_n + \omega_m)t}{\beta n \; \beta m \quad \beta n + m} \qquad 6.10$$

where $\beta_j = [(\omega_o^1)^2 + i\gamma\omega_j - \omega_j^2]$.

This analysis, within the limitations of being a) one-dimensional, b) dealing with anharmonicity of only second order and c) assuming the presence of only one type resonant oscillator, serves to indicate the type of polarisation terms to be expected. Thus we may write the first-order susceptibility in the form

$$\chi^{(1)}(\omega_n) = \frac{Ne^2}{m} \cdot \frac{1}{(\omega_o^1)^2 + i\gamma\omega_n - \omega_n^2} \qquad 6.11$$

The corresponding second-order susceptibility will, from (6.10) and (6.11), take the form

$$\chi^{(2)}(\omega_n,\omega_m) = \frac{\xi m}{N^2 e^3} [\chi^{(1)}(\omega_n) \, \chi^{(1)}(\omega_m) \, \chi^{(1)}\omega_{n+m})]. \qquad 6.12$$

In the one-dimensional case considered above, the susceptibilities are scalar quantities. In three dimensions, the susceptibilities, which link the vector quantities $\underline{P}$ and $\underline{E}$, are third-rank tensors, so that

$$P_u(\omega_{n+m}) = \sum_{vw} \sum_{nm} \chi_{uvw}(\omega_n,\omega_m,\omega_{n+m}) E_u(\omega_n) E(\omega_m)$$

$$\exp i(\omega_n + \omega_m)t \qquad 6.13$$

where $\omega_{n+m} = \omega_n + \omega_m$

If we assume that the only frequencies present are ω_1,ω_2 and the sum frequency $\omega_3 = \omega_1 + \omega_2$, then the (x, y, z) components of the polarisation at frequency ω_1 are given from equation (6.13) by

$$P_u(\omega_1) = \sum_{vw} \left[\left\{ \chi_{uvw}(\omega_1, -\omega_2,\omega_3) \, E_v(-\omega_2) E_w(\omega_3) \right. \right.$$

$$+ \chi_{uvw}(\omega_1,\omega_3,-\omega_2) E_v(\omega_3) E_w(-\omega_2) \bigg\} \, e^{i\omega_1 t}$$

$$+ \left\{ \chi_{uvw}(-\omega_1, \omega_2, -\omega_3) E_v(\omega_2) E_w(-\omega_3) \right.$$

$$+ \chi_{uvw}(-\omega_1,-\omega_3, \omega_2) E_v(-\omega_3) E_w(\omega_2) \bigg\} \, e^{-i\omega_1 t} \Bigg]$$

$$6.14$$

where v,w take all possible values of (x,y,z). Corresponding expressions for $P_u(\omega_2)$ and $P_u(\omega_3)$ are readily obtained from equation (6.13). If the frequencies of the waves present are well away from the resonance value ω_0^1, then equation (6.12) shows that $\chi^{(1)}(\omega_1, \omega_2) \simeq \chi^{(2)}(-\omega_1,\omega_2)$, which reduces the number of χ terms to a maximum possible of 81. In fact, equation (6.12) also shows that

$$\chi_{uvw}(\omega_1,-\omega_2,\omega_3) = \chi_{uvw}(\omega_2,\omega_3,-\omega_1) = \chi_{wuv}(\omega_3,\omega_1,\omega_2) \qquad 6.15$$

so that there are in fact only 27 independent terms. In fact since the order in which the ω_1,ω_2 fields are applied makes no difference, the number of independent terms reduces to 18. This number will in general be further reduced in a manner determined by crystal symmetry.

It has been established by Miller that the second-order susceptibility is related to first-order values by the relation

$$\chi^2_{uvw}(\omega_1,\omega_2,\omega_3) = [\chi^{(1)}(\omega_1) \chi_{vv}^{(2)} (\omega_2) \chi_{ww}^{(2)} (\omega_3)] \Delta_{uvw} \qquad 6.16$$

where Δ_{uvw} is approximately constant for a wide range of materials.

6.3   Second Harmonic Generation (SHG)

With $\omega_2 = \omega_1$ in the previous section, the non-linearity of the system clearly leads to second harmonic generation. As in the case of electro-optic tensor (section 4), a shortened notation may be employed for the susceptibility terms according to the following scheme:

$$d_{ixx} \rightarrow d_{i1} \qquad\qquad\qquad d_{iyz} \rightarrow d_{i4}$$

$$d_{iyy} \rightarrow d_{i2} \qquad\qquad\qquad d_{izx} \rightarrow d_{i5}$$

$$d_{izz} \rightarrow d_{i3} \qquad\qquad\qquad d_{ixy} \rightarrow d_{i6}$$

With this notation, the polarisation is related to the applied field by

$$\begin{pmatrix} P_x \\ P_y \\ P_z \end{pmatrix} = \begin{pmatrix} d_{11} & d_{12} & d_{13} & d_{14} & d_{15} & d_{16} \\ d_{21} & d_{22} & d_{23} & d_{24} & d_{25} & d_{26} \\ d_{31} & d_{32} & d_{33} & d_{34} & d_{35} & d_{36} \end{pmatrix} \begin{pmatrix} E_x^2 \\ E_y^2 \\ E_z^2 \\ 2E_yE_z \\ 2E_zE_x \\ 2E_xE_y \end{pmatrix} \qquad 6.18$$

Many of the terms d_{ij} vanish in the face of crystal symmetries. Thus for crystal classes $\bar{4}3m$ and 23, the only non-vanishing terms are $d_{14} = d_{25} = d_{36}$. If the dispersion of the non-linear coefficients is negligible, then further reduction occurs in the number of independent components (Kleinman's symmetry conditions). Thus although the symmetry of class 222 implies three non-vanishing elements d_{14}, d_{25} and d_{36}, the effect of Kleinman symmetry is that these are in fact equal. Similarly for class mm2, the structural symmetry leads to d_{15}, d_{24}, d_{31}, d_{32} and d_{33}. The Kleinman condition makes $d_{31} = d_{15}$ and $d_{32} = d_{24}$. In the triple-suffix notation d_{ijk}, the Kleinman condition states that all coefficients with permuted i,j,k, are equal. The full set of equalities is thus:

$$\left. \begin{array}{lll} d_{12} = d_{26}, & d_{13} = d_{35}, & d_{14} = d_{25} = d_{36} \\[1.5ex] d_{21} = d_{16}, & d_{22} = d_{15}, & d_{32} = d_{24} \\[1.5ex] & d_{22} = d_{34} & \end{array} \right\} \qquad 6.19$$

A requirement for efficient S.H.G. is that the velocities of fundamental and harmonic waves in a given direction shall be equal. In a birefringent medium, this may be possible. The birefringence of certain materials depends on temperature in such a way that a suitable matching ('phase-matching') may be achieved by adjustment of the temperature. Thus for lithium niobate, which is uniaxial, the velocities of the ordinary ray for a wavelength of 1.064 μm may be made to match that of the extraordinary ray for 0.532 μm for a direction perpendicular to the optic axis of the crystal by adjusting the temperature to about 60°C, the precise temperature depending rather sensitively on the stoichiometry of the crystal.

For this simple situation, we may use a one-dimensional argument. From Maxwell's equations, and taking $\mu_r = 1$, we have

$$\nabla^2 \underline{E} = \mu_o \epsilon_o \frac{\partial^2}{\partial t^2} (\epsilon \underline{E}) - \mu_o \epsilon_o \frac{\partial^2 \underline{P}}{\partial t^2} \qquad 6.20$$

If we impress signals of frequencies ω_1, ω_2, such that

$$E_1(z,t) = E_1(z) \exp[-i(\omega_1 t - k_1 z)]$$
$$E_2(z,t) = E_2(z) \exp[-i(\omega_2 t - k_2 z)] \qquad 6.21$$

then form a suitably simplified from of equation (6.14), we have, for the induced polarisations,

$$P_1(z,t) = \chi E_1{}^* E_2 \exp[-i (\omega_2 - \omega_1) t - (k_2 - k_1) z]$$
$$P_2(z,t) = \chi E_1{}^2 \exp[-i(2\omega_1 t - 2k_1 z)] \qquad 6.22$$

From 6.20 - and 6.22 and neglecting terms in $\dfrac{d^2 E}{dz^2}$ compared with $\dfrac{dE}{dz}$, we obtain for the case $\omega_2 = 2\omega_1$

$$\frac{dE_1}{dz} = -\frac{i\mu_o \epsilon_o \omega_1{}^2}{k_1} \chi E_1{}^* E_2 \exp[-i (2k_1 - k_2) z] \qquad 6.23$$

$$\frac{dE_2}{dz} = - \frac{2i \, \mu_o \epsilon_o \omega_1{}^2}{k_2} \chi E_1{}^2 \exp \left[i \left(2k_1 - k_2 \right) z \right] \qquad 6.24$$

If we restrict consideration to the case where the amplitude of the second harmonic is small with that of the fundamental, then equation (6.24) can be integrated to give, for a path of length ℓ in the medium,

$$E_2(\ell) = \frac{- 2\mu_o \epsilon_o \omega_1{}^2 \, \chi E_1{}^2}{k_2} \left(\frac{e^{i\ell\Delta k} - 1}{\Delta k} \right) \qquad 6.25$$

where $\Delta k \equiv 2k_1 - k_2$.

The Poynting vector $S \equiv cn|E|^2$, so the second-harmonic flux $S(2\omega_1)$ is given in terms of $S(\omega_1)$ by

$$S(2\omega_1) = \frac{4 \, \omega_1{}^4 \, n(2 \, \omega_1) \, \chi^2 \, \ell^2 \, \text{sinc}^2 \left(\frac{\ell\Delta k}{2} \right)}{c^5 k_2{}^2 n^2 (\omega_1)} \, S^2(\omega_1) \qquad 6.26$$

In the presence of exact phase-matching, $\Delta k = 0$ and $\text{sinc}^2(\frac{\ell\Delta k}{2}) = 1$, so that (6.26) reduces to

$$S(2\omega_1) = \frac{4 \pi^2 \, n(2\omega_1) \, \chi^2 \, \ell^2 \, S^2 (\omega_1)}{c\lambda^2} \qquad 6.27$$

We may remove the small-signal restriction as follows. Writing equations (6.23) and (6.24) as

$$\frac{dE_1}{dz} = - i\alpha E_1{}^* E_2 e^{- i\Delta k.z} \qquad 6.28$$

$$\frac{dE_2}{dz} = - i\beta E_1{}^2 e^{i\Delta k.z} \qquad 6.29$$

where $\quad \alpha \equiv \dfrac{\mu_o \epsilon_o \omega_1{}^2 \chi}{k_1}$ and $\beta \equiv \dfrac{2\mu_o \epsilon_o \omega_1{}^2 \chi}{k_2}$

and putting $E = \tfrac{1}{2} E \, e^{i\phi}$, we obtain

$$\frac{dE_1}{dz} = -\frac{\alpha}{2} E 1 \, E 2 \sin \theta \qquad 6.30$$

$$\frac{dE_2}{dz} = \frac{\beta}{2} E_1{}^2 \sin\theta \qquad\qquad 6.31$$

and $$\frac{d\theta}{dz} = \Delta k - (\alpha E_2 - \frac{\beta E_1{}^2}{2E_2}) \cos\theta \qquad\qquad 6.32$$

where $\theta \equiv 2\phi_1 - \phi_2 + \Delta k.z$

If we consider the case of perfect phase-matching ($\Delta k = 0$) and one where a wave of frequency ω_1 is incident at $z = 0$, with no second harmonic power at this point ($E_2(o) = 0$), then with

$$p \equiv \frac{E_1(z)}{E_1(o)} \quad , \quad q = \frac{E_2(z)}{E_1(o)}$$

and $$r = \frac{\alpha}{2} E_1(o)z \; (= \frac{\beta}{2} E_1(o) z \text{ for } 2k_1 = k_2),$$

$$\frac{dp}{dr} = - pq \sin\theta \qquad\qquad 6.33$$

$$\frac{dq}{dr} = p^2 \sin\theta \qquad\qquad 6.34$$

From equation (6.32), we have, for $\Delta k = 0$ and $\alpha = \beta$,

$$\frac{d\theta}{dr} = (\cot\theta) \cdot \frac{d}{dr} (p^2 q) \qquad\qquad 6.35$$

which may be integrated to give

$$p^2 q \cos\theta = 0 \text{ (since } q(o) = 0) \qquad\qquad 6.36$$

Since the energy flux in the wavefield is proportional to $E_j{}^2$ and since no losses are assumed, then

$$E_1{}^2(z) + E_2{}^2(z) = E_1{}^2(o) \qquad\qquad 6.37$$

or $$p^2 + q^2 = 1 \qquad\qquad 6.38$$

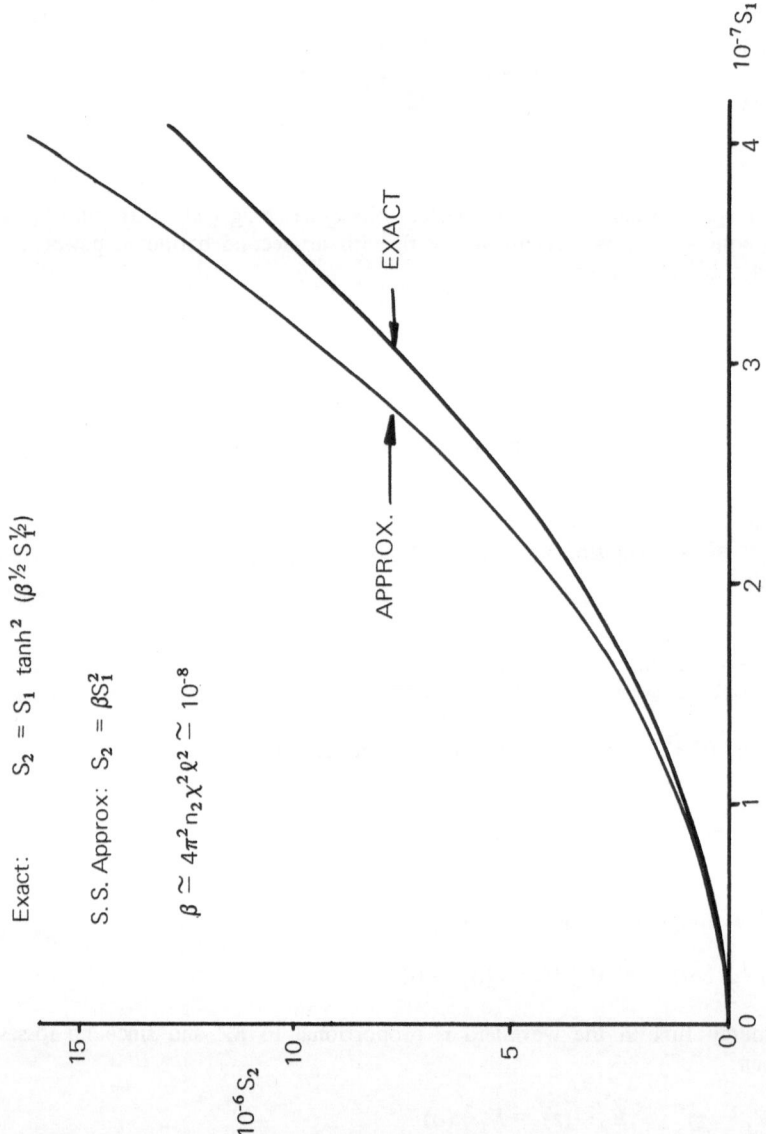

Exact: $S_2 = S_1 \tanh^2 (\beta^{1/2} S_1^{1/2})$

S. S. Approx: $S_2 = \beta S_1^2$

$\beta \simeq 4\pi^2 n_2 \chi^2 \ell^2 \simeq 10^{-8}$

Fig. 5. Second harmonic power as function of incident power: upper curve — small-signal approximation; lower curve — exact treatment.

From (6.34), (6.36) and (6.38),

$$\frac{dq}{dr} = \pm (1 - q^2) \qquad\qquad 6.39$$

or, since $q(o) = 0$, $q = \tanh r$, which in terms of the incident field $\xi_1(o)$ may be written, from the definition of q,

$$E_2(\ell) = E_1(o) \tanh (\ell / \ell_s) \qquad\qquad 6.40$$

ℓ_s is a measure of the distance in which a substantial conversion of fundamental to second harmonic occurs and is given by

$$\ell_s = \frac{2}{\alpha E_1(o)} = \frac{2k_1}{E_1(o)\mu_o \epsilon_o \omega_1^2 \chi} \qquad\qquad 6.41$$

This may more usefully be expressed in terms of the parameters of the medium and the incident power S_o (watts m^{-2}) in the form

$$\ell_s = \frac{nc^{\frac{1}{2}}\lambda_o}{\chi S_o^{\frac{1}{2}}} \qquad\qquad 6.42$$

where λ_o is the vacuum wavelength of the fundamental.

It may be noted that, for small values of ℓ, equation (6.40) reduces to

$$S(2\omega_1) = \frac{cn(2\omega_1) E_2^2(\ell)}{4} = \frac{4\pi^2 n (2\omega_1) \chi^2 \ell^2 S^2 (\omega_1)}{c\lambda^2} \qquad\qquad 6.43$$

in agreement with equation (6.27) (Note that, from the definition of E, we have $S = cn|E|^2 = \frac{cnE^2}{4}$).

The exact and small-signal cases are illustrated in Fig. 5

The analysis of SHG, and other non-linear effects in cases where phase-matching cannot be achieved is considerably more cumbersome than that presented above, which is intended only to outline the phenomenon. The general case is dealt with by Armstrong et al. (1962).

RECOMMENDED READING

Coherent Light, A. F. Harvey; Wiley-Interscience; 1970

Applied Nonlinear Optics, F. Zernicke and J. E. Midwinter; Wiley-Interscience; 1973

Armstrong, J. A., Bloembergen, N., Ducuing, J. and Persham, P. S., 1962, Phys Rev, $\underline{127}$, 1918-1939

RECENT DEVELOPMENTS IN HOLOGRAM

Radar and Sonar

W. E. Kock

Acting Director
The Herman Schneider Laboratory of Basic and Applied Sciences Research
University of Cincinnati
Cincinnati, Ohio (U. S. A.)

ABSTRACT:
Following a review of the synthetic aperture (hologram) radar concept, several new developments are described. The first is the acquiring of range information by holographic means alone (that is, without the use of pulses). This process acquires two-dimensional data from a one-dimensional record, comparable to the acquiring, in a two-dimensional hologram, of three-dimensional information. The second development utilizes broad band, incoherent signals for the illuminating and reference waves in a bistatic radar or sonar system, with narrow band filtering procedures for generating a plurality of one-dimensional zone plates. The third development permits the maximum range of synthetic aperture systems to be extended by many times. The paper concludes with a brief note discussing the parallel processing aspects of the synthetic aperture technique.

HOLOGRAMS AND SYNTHETIC APERTURE SYSTEMS

One of the most extensive uses of non-optical holograms which has occurred in the microwave radar field is in synthetic aperture radar. Such radars, developed in the 1950's and 1960's (1), are now recognized as being a form of holography (2, 3). Similar hologram concepts are beginning to be examined for ultrasonic use also.

In a synthetic aperture radar system, an aircraft moving along a very straight path continually emits successive microwave pulses. The frequency of the microwave signal is very constant; the signal remains "coherent" with itself for very long periods. During these periods the aircraft may have travelled several thousand feet but, because the signals are coherent, all the many echoes which return during this period can be processed as though a single antenna as long as the flight path had been used. The effective antenna length is thus quite large. This large "synthetic" aperture has very high resolving power, even at microwaves, enabling the radar to present extremely fine detail.

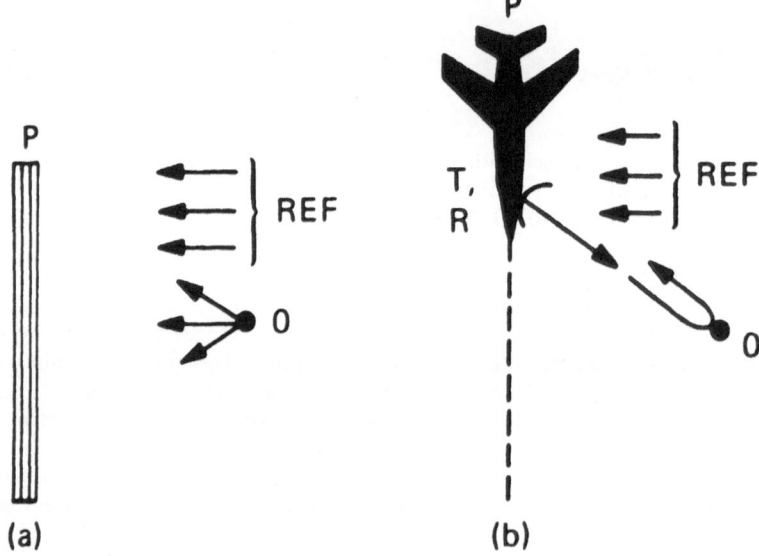

Fig. 1. In a synthetic aperture radar (right), echoes from the reflecting point are received over a considerable interval and assembled by a holographic method that creates interference patterns between the return signal and a reference sampled from the transmitted signal. The result is a synthetic aperture, high-resolution antenna whose length is that of the flight path. The similarity to the hologram process (left) is evident.

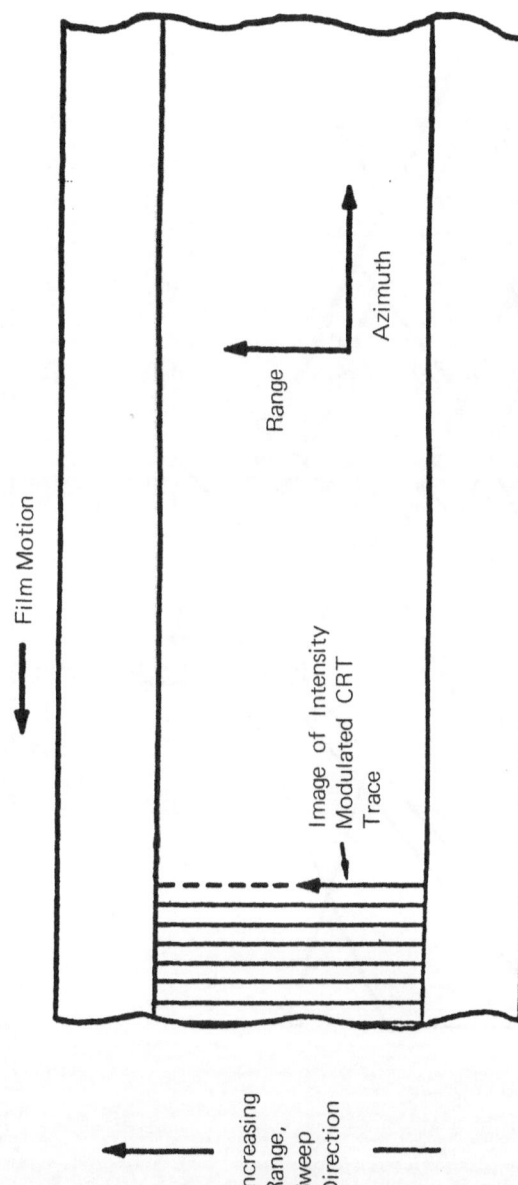

Fig. 2. Photographic recording arrangement for forming synthetic aperture radar records.

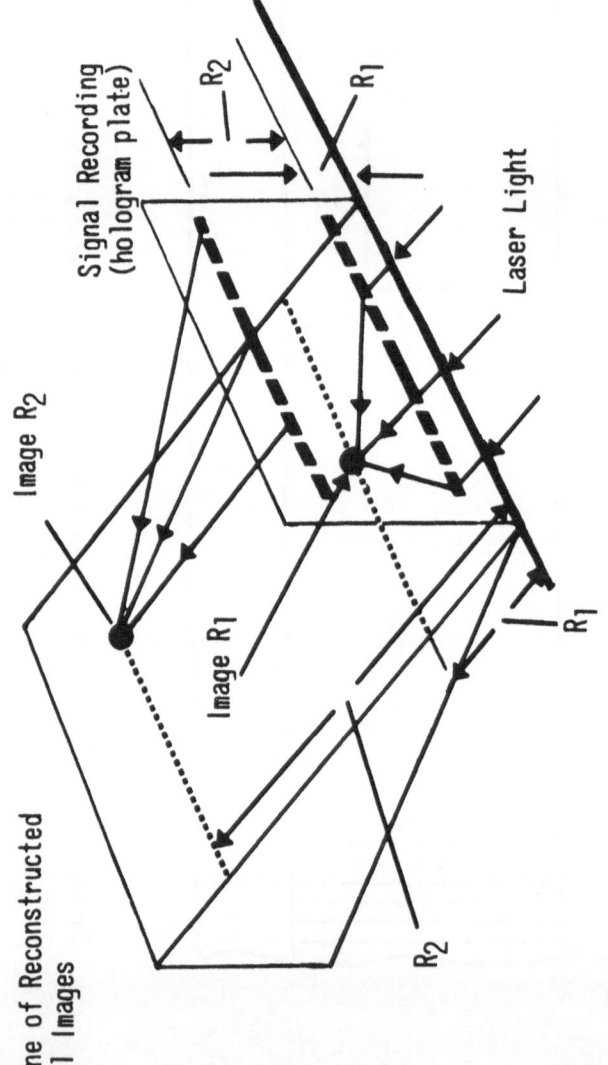

Fig. 3. Photographically recorded one-dimensional zone plates permit coherent laser reconstruction of real images of two reflecting points displaced appreciably in range and slightly in azimuth.

The photographic record of the echoes received by such a coherent radar is the hologram; through the process of synchronous detection, the microwave generator which provides the illuminating signal also provides a reference wave. The reflected signals received along the flight path are made to interfere with this reference signal, and the complex interference pattern thereby generated and photographically recorded is the hologram.

The method of operation is sketched at the right in Fig. 1. For simplicity, only one reflecting point is shown. Waves returning from this point have spherical wave fronts whereas the oscillator reference signal acts like a set of plane waves perpendicular to the path of the airplane. The similarity of this process to the one on the left, where waves from a point source, 0, are caused to interfere with a set of plane reference waves, is evident. The received signal is combined with the coherent reference signal and caused to intensity modulate a cathode-ray tube trace as shown in Fig. 2. Each vertical line in that figure thus plots signals received from all range points, with the points at greater range being recorded near the top of the vertical trace. As the airplane moves along and new pulses are emitted, the film is indexed to record a new set of returns.

For the case of only one reflecting point at fixed range, the upward moving cathode ray beam would, for every pulse, be modulated only at that one point in range, and the result would be a single horizontal line of recorded echoes. But this line is not continuous. The returning waves are circular and as the slant azimuth range from the aircraft to the reflecting point changes, the combination of return and reference waves produces successive constructive and destructive interference. At the greater slant angles, this succession of in-phase and out-of-phase conditions occurs rapidly, whereas when the aircraft is practically abreast of the point, it occurs slowly. The resulting record is a one-dimensional zone plate hologram. Illuminated by laser light, it would reconstruct the reflecting point.

The multitude of reflecting objects in the terrain flown over by the aircraft are reconstructed by illuminating the hologram with a laser as shown in Fig. 3. Indicated in the figure are two reflecting points which are displaced appreciably in range and slightly in azimuth. All reconstructed images fall on a plane. The tilt of this plane is determined by the amount of radar vertical tilt.

A typical microwave hologram as generated by a side-looking radar is shown in Fig. 4. In this figure, a prominent one-dimensional zone plate is seen near the bottom of the blank areas. The processed radar records (the holograms) yield photos of extremely good detail (Fig. 5.).

When synthetic aperture radar was first described, many could not understand how its high resolution could be maintained for nearby objects. Because of the great length of the synthetic aperture antenna, reflecting objects are in the near field, not in the distant, or Fraunhofer region, where most radar antennas operate. For a near-field reflecting point object, radars having large aperture antennas can only achieve maximum efficiency and resolution when the phases of their receiving elements are

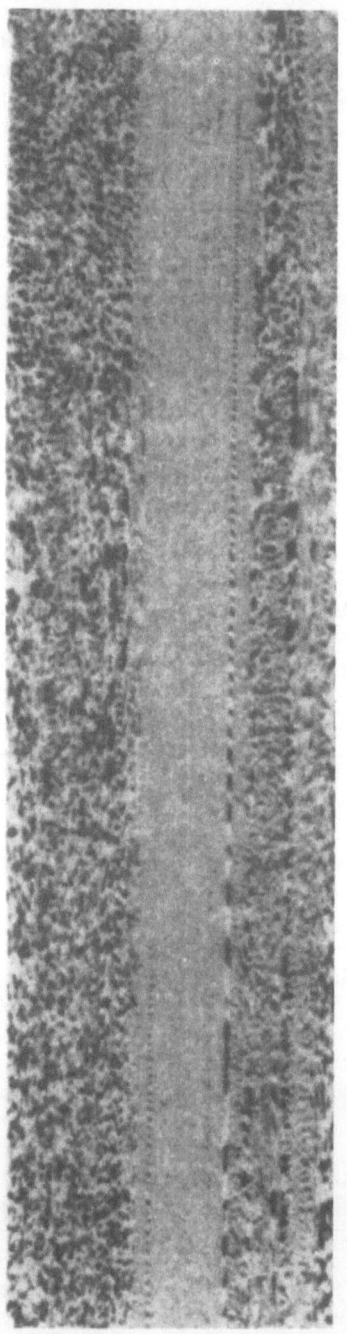

Fig. 4. A hologram record of a synthetic-aperture radar exhibits a prominent one-dimensional zone-plate signature for a particularly distinct reflecting object (lower portion of blank area).

Fig. 5. A synthetic aperture radar "photograph" (a hologram reconstruction) of the city of Phoenix, Arizona, U.S.A. (at the middle right). The block patterns at the top and at the left are irrigated farms. The geological features are clearly seen at the top right, at the bottom right, and at the bottom left. The white area is blocked out for security reasons. (courtesy Goodyear Aerospace Corporation).

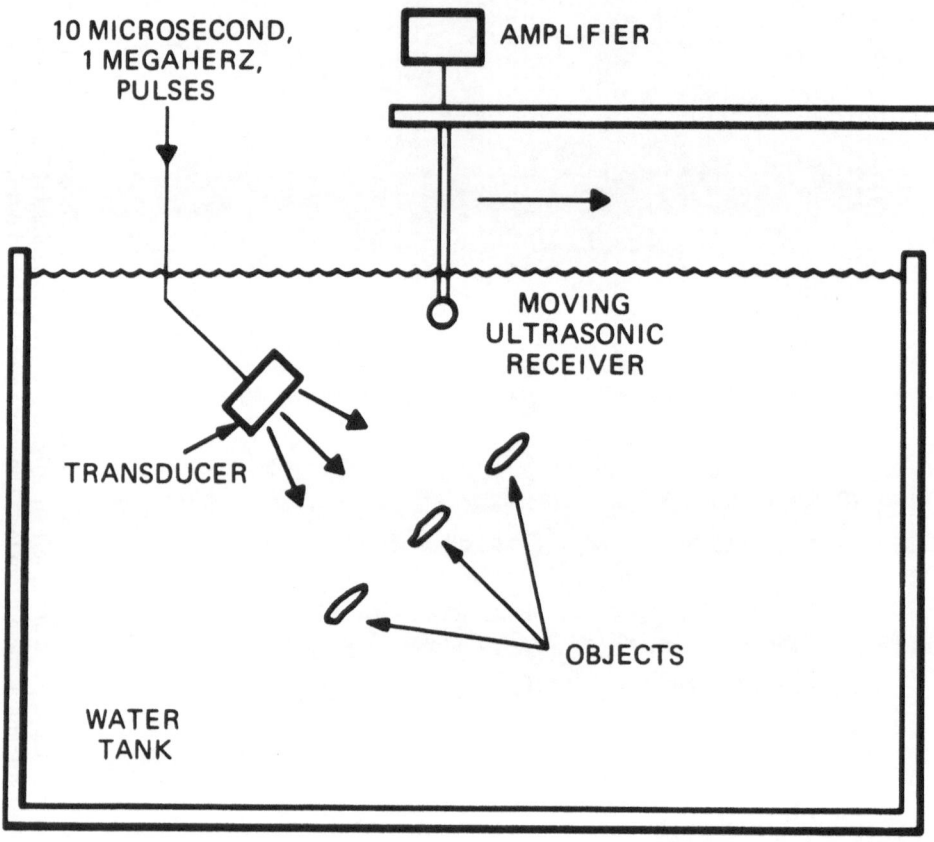

Fig. 6. Generating one-dimensional synthetic-aperture zone plates with a moving ultrasonic receiver. The reflected pulse signals from three objects, when combined with a reference wave (by synchronous detection), are three zone-plates. F. Pekau and Diehl, Reference (5).

adjusted so as to correspond to arcs of circles centered on these other points. How the mysterious optical processor could cause all points, at any distance, to be in sharp focus, was beyond comprehension to many.

But when it is recognized that the radar record is a hologram, the solution becomes obvious. In holography, each small light-reflecting point generates its own zone plate, and each zone plate later causes coherent light to be reconstructed exactly at the point in space from whence the light emanated. Similarly, a side-looking radar captures photographically the curved wave fronts emanating from a reflecting point by combining them with a reference wave in a one-dimensional zone plate. Later, as in an optical hologram, coherent reconstructing laser light acquires, through diffraction by zone plate, the properly curved wave fronts to focus light at points corresponding to the reflecting points in the original landscape. Just as an optical hologram causes each point of a three-dimensional scene to be brought into sharp focus no matter what its distance from the photographic plate, so the microwave hologram (the radar record) is responsible for the good focus of all its reconstructed points.

The use of the synthetic aperture concept in ultrasonic applications is described in a U.S. patent applied for in March, 1967, and called "Synthetic Aperture Ultrasonic Imaging Systems" (4). Two procedures are described, both aimed at medical ultrasonic applications for the "examination of the interior of living bodies". The first is exactly that of the radar form, where an ultrasonic transducer is moved (scanned) along a straight line, along a liquid reservoir, with the liquid in contact with the body surface under examination. The second procedure utilizes a very large number of fixed transducers positioned along the line of motion of the moving transducer of the other procedure, and energized in succession so as to stimulate the moving transducer. This moving of both transducer and receiver causes the effective length of the array of transducers to be doubled and thus provides added information (more hologram fringes are generated).

More recently, experiments using the ultrasonic synthetic aperture procedure for true sonar use were described, with very excellent records resulting (5). Figure 6 shows the arrangement of equipment used in these tests; the transmitter radiated 10 microsecond pulses of 1 MHz sound waves. It is seen that only the receiving transducer was moved, with the round trip distance to each of the three reflecting objects from transmitter and receiver being sufficiently different to permit the one-dimensional zone plates to be adequately separated in the holographic record. For these tests, the reference waves were supplied electronically from the stabilized oscillator directly to the receiver, which also supplied energy (amplified) to the transmitting transducer. It was expected that three one-dimensional zone plates (one for each reflecting object) would be generated, all adequately separated to permit reconstruction, from the zone plates, of the three individual objects. Fig. 7 shows the very clear one-dimensional zone-plates which resulted. The similarity of these zone-plates to the radar one of Fig. 4 is evident. It is seen that the left-hand zone plate has its central portion slightly above center of the photo, whereas the middle zone plate has its slightly below, and the right hand one has its below the bottom of the photo. The right hand portion of Fig. 7 shows the reconstruction (by laser light) of the three reflecting objects (the light areas at the lower left are artifacts). One of the two co-investigators

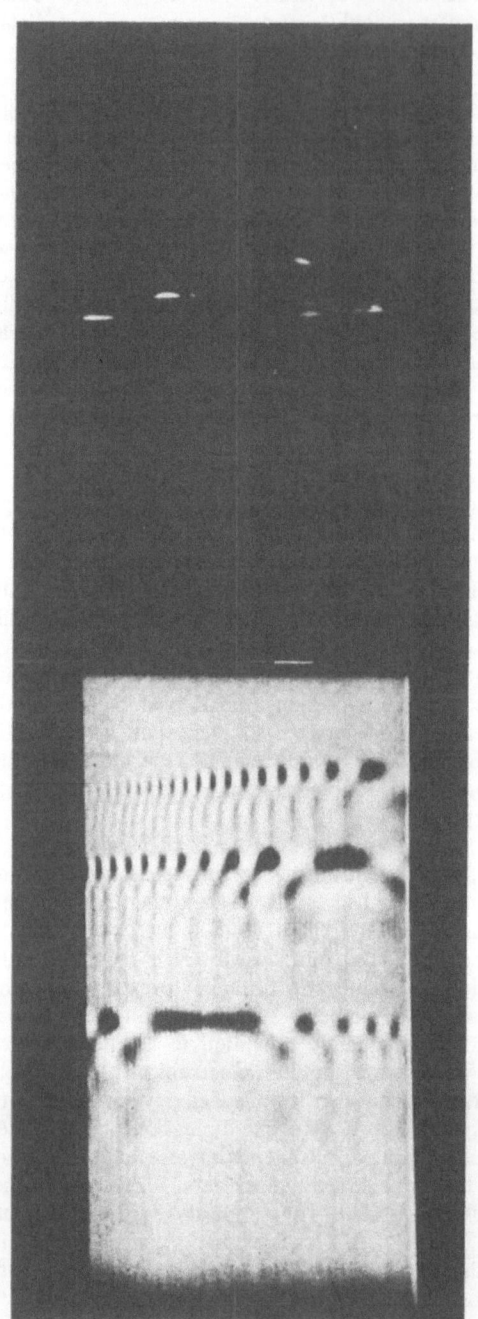

Fig. 7. At the left are the three one-dimensional zone plates as generated in Fig. 12. At the right are the laser reconstructed images of the three objects (the lower left white marks are artifacts). (F. Pekau and Diehl).

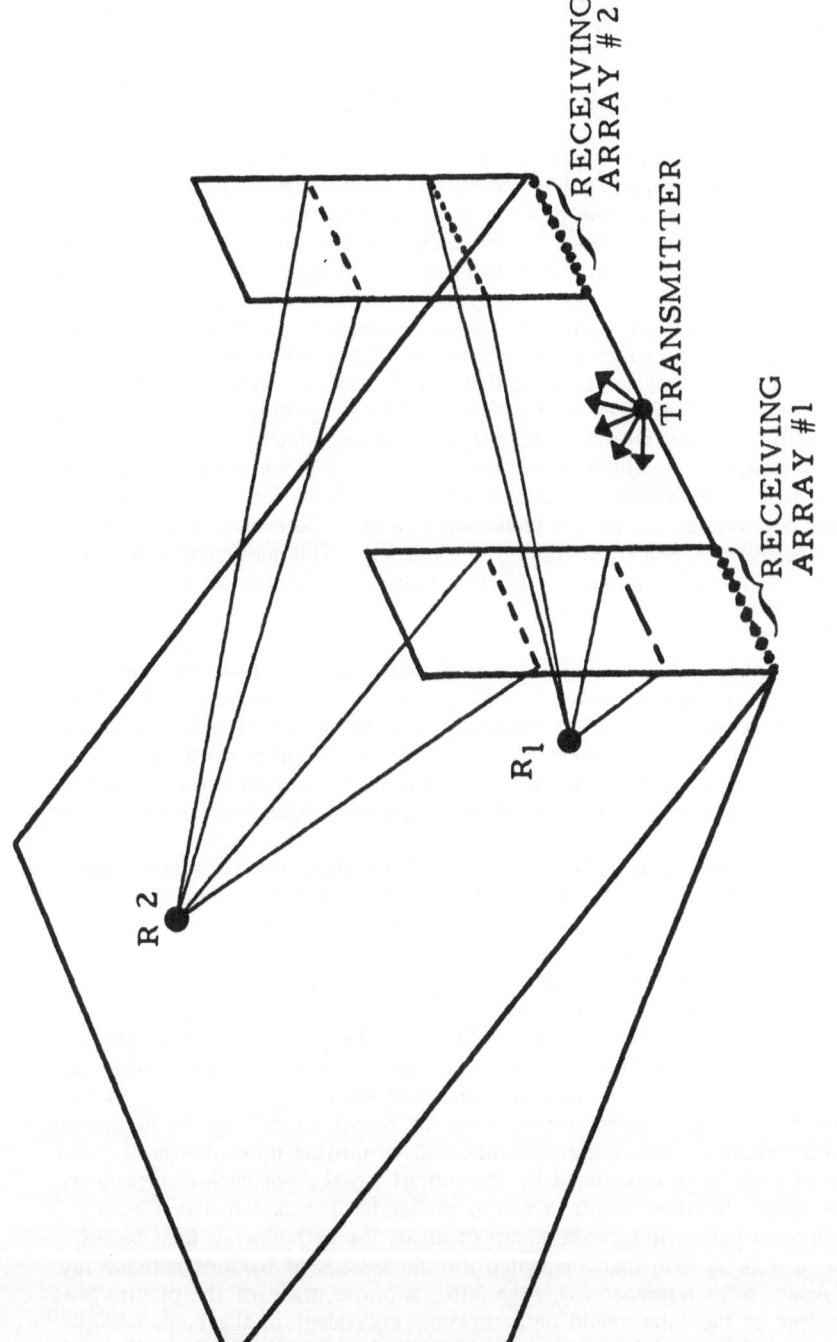

Fig. 8. A stationary line array, when employed as a holographic radar receiving array, can employ portions of the array.

in this work (Dr. F. Pekau), had earlier been associated with the Bendix Research Laboratories.

RADAR RANGE ACQUISITION BY HOLOGRAPHY

Both the usual radar systems, and those utilizing the synthetic aperture technique just described, employ pulses for measuring the range to a target, as determined by the time taken by the pulse in travelling from the transmitter to the target and back to the receiver. Once it was recognized that the synthetic aperture procedure is a form of holography (2), other forms of hologram radars became apparent, including end-fire (synthetic gain) systems (6), bistatic systems (7) and various stationary extensions of moving synthetic aperture systems (3,8). It was in the discussions of such stationary hologram radar systems that the concept of range acquisition by hologram means was first described, and, these descriptions were referred to in a recent publication (9). Also, recently, the first published experimental data demonstrating validity of the concept was reported (9). In these experiments, a situation was simulated on a computer in which an S-band (10 centimeter) wavelenght continuous wave (not pulsed), radar, having an antenna aperture of 10 meters, was found to exhibit a huge response at the proper (zoneplate) range of 28 meters and practically zero response at other ranges (from zero to 200 meters). This particular hologram radar is being constructed "for purpose of measuring the thickness of ice layers" (quoted from reference 9).

The sketch in Figure 8 shows a stationary radar based on the synthetic aperture concept. In holography, even a small portion of the hologram is able to reconstruct the full image. Similarly, in stationary coherent sonars or radars, retaining only the end sections of along linear array (thereby maintaining maximum resolution) can be satisfactory in many situations. Such a use in a linear array is shown in the figure here, the transmitter is placed at the unused center portion of the array.

In the pulsed form of this system, the transmitter would radiate (from Reference 3, page 335) a "short pulse of coherent microwaves following which all (receiver) units would, as in an ordinary radar, act as receiving elements. A certain amount of the coherent microwave signal would continually be fed to all of the elements, thereby providing the holographic reference signal. The returning, reflected, echo signals would interfere at each receiver this reference wave, so that along the entire array, a wave interference pattern would exist. This pattern could be photographically recorded, for example through the use of a cathode ray tube having, instead of a single beam, a large number of beams (the number corresponding to the number of elements in the receiver array), with the beams all striking the luminous face of the tube along a single horizontal line, and all moving upward together, and the intensity of each being modulated by the output signal from each elemental receiver of the array". In other words, a record similar to that sketched in Figure 2 would be generated with one upward sweep of all of the cathode ray tube beams, instead of, as in Fig. 2, with many repeated upward sweeps of the one cathode ray tube beam. Again, from reference (3), page 336, "a photo made of the picture generated on the face of the tube would be completely equivalent to the radar record" (e.g. Fig. 4), since "all reflecting objects would similarly generate one-dimensional zone plate interference patterns. The holographic reconstruction would similarly provide a picture of the area located in the field of the radar". In Fig. 8 the two vertical rectangles indicate the zone plate recordings of two reflecting points R_1 and R_2, comparable to the two reflecting points in Fig. 3.

As has been noted (3), when the individual elemental receivers in a holographic line array (such as in Fig. 8) possess an aperture, of one wavelength or less, the range and azimuth resolutions of the array become equal. From reference (3), page 338, "under these conditions good range resolution would be acquirable without the use of pulses, i.e., through the action of the superimposed zone plates alone." In this case (for example in Figure 8), "no vertical of the large number of cathode ray tube spots would be required. They would remain stationary, generating a single, horizontal, line-pattern. The procedure would result in all of the (horizontal) one-dimensional zone plates in the radar record being superimposed to form one horizontal line. (In the usual radar record, superposition of the one dimensional zone plates occurs only for reflectors at the same range; in this situation superposition occurs for all reflecting points). As in a true hologram, however, (for which this superposition of zone plates of reflecting objects at all ranges also occurs), reconstruction of the original scene is not hampered by this superposition."

It is interesting to compare the data acquisition of such a continuous wave stationary synthetic aperture (hologram) line array radar with the data acquisition in a hologram. As the author has discussed (reference 11, page 76), in the early days of laser holography it was difficult for many to understand how a single, two dimensional, photographic plate, or a light beam having two degrees of freedom, could carry information about a three dimensional object (described by three degrees of freedom). In the continuous wave radar case described above, a one-dimensional line of superimposed zone plates carries full information about a two-dimensional area, just as the (synthetic) single line array of the aircraft captured the highly detailed information later to be reconstructed as the highly detailed photo of Fig. 5.

In the recent 1973 reference (10), there is some discussion of the "laborious task" of the adjustment of the phase shifters to obtain focussing of the receiving array. A similar focussing situation in a near field sonar is discussed in two of the author's forthcoming publications (12, 13), in which a 400 kHz sonar, using a 29 element, 3 meter long, linear array is employed. To properly focus this (near-field) receiving array at all ranges a 24 step electronic sequence is performed in which the required phase adjustments for each 10 meter range are provided so as to properly focus the array inward to each (on-axis) near-field point in the object plane. These phase adjustments are all different for each of the 29 receiver elements and for each of 24 range steps. It is pointed out (12, 13) that one of the significant advantages of the hologram procedure in near-field sonar (or radar) is its ability to automatically provide focussing without the need to phasing procedures. The focussing is provided, as discussed above, through the zone-plate forming action.

HOLOGRAM RADAR USING INCOHERENT ILLUMINATION

In the usual procedure for making a hologram, and in hologram radar

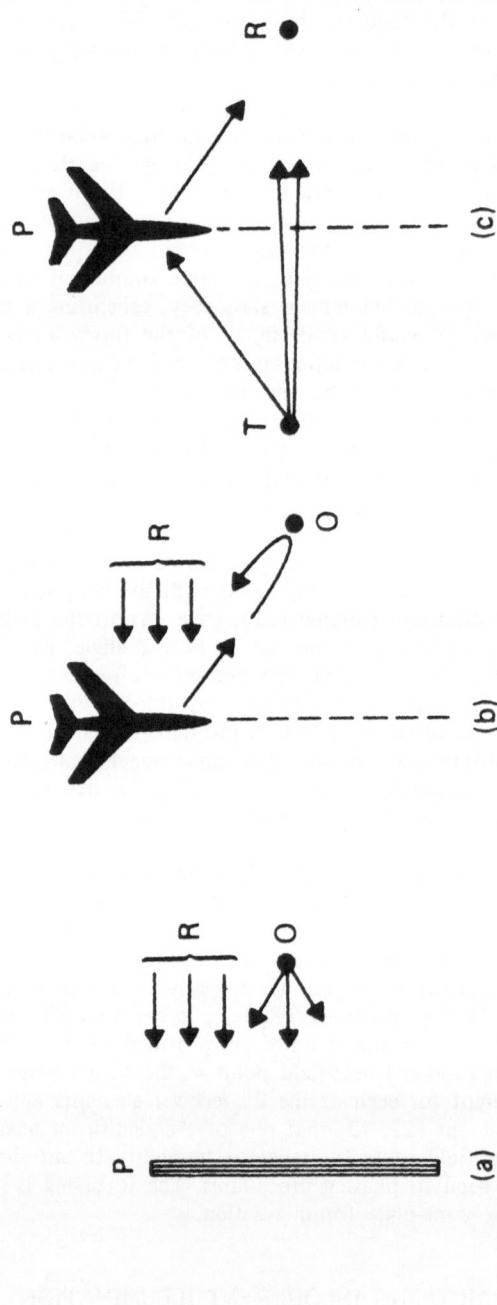

Fig. 9. Zone-plate signals, as generated by a synthetic aperture system (left) are also generated when a target passes between a bistatic arrangement (either radar or sonar).

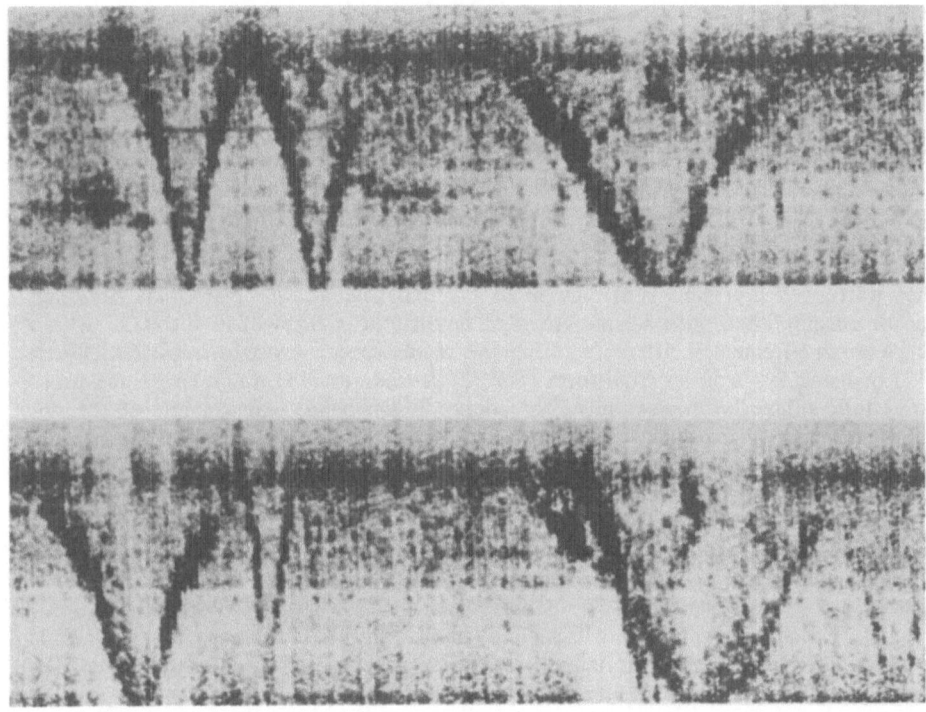

Fig. 10. Time-frequency record of the interference effects caused by aircraft flying over a transmission path illustrates principle underlying holographically generated bistatic radar or sonar zone plate. Holographic techniques can be used to enhance weak signals such as those shown on the bottom right of the record.

techniques, highly coherent, single-frequency waves are used for illuminating the object and for the reference wave. In some radar and sonar applications, however, a broad bandwidth transmitted signal is more desirable than a single frequency one. Noise-like signals are less easily detected and are more difficult to jam or to home on. They also permit higher power to be radiated when power limitations exist in any of the transmitter components. This, is for example, the usual reason for using wide-band "chirp" signals. Recently, a procedure was described (14) in which a random noise source is used in place of the coherent waves normally used in hologram radar or sonar systems. It has been noted (7) that in a continuous wave bistatic radar or sonar system (Fig. 9 (c)), a target, P, crossing the line between transmitter and receiver at uniform speed, generates a signal (as in Fig. 10) which is a one-dimensional zone plate. The similarity of Fig. 9 (c), to (a) and (b), (as in Fig. 1) is evident. If, in Fig. 9, the coherent transmitter signal T is replaced with a noise signal, the unprocessed combined signal at R would obviously not exhibit such a pattern. However, when this combined signal is given a time-frequency, narrow-band analysis, its presentation then portrays a multiplicity of contiguous one-dimensional zone plates. The filtering can be accomplished with a large group of contiguous, narrow-band filters, or with a single varying-frequency filter, i.e., using the visible speech procedures of R.K. Potter (15) or using fast Fourier Transform (F.F.T.) procedures (16), (17). (All three procedures fully utilize, of course, all of the energy in the band). An example of the second procedure is given in the acoustic spectrogram of Fig. 11. For the photo a noise source was combined with a delayed replica of itself, of opposite polarity, with the delay (positive and negative), being varied at a constant rate. It is seen that the interference fringes generated by the higher frequencies of the noise signal are more closely spaced than those at lower frequencies. Fig. 12 portrays an F.F.T. spectrogram of a changing underwater acoustic interference pattern (taken from reference (17)).

Had the geometry of Fig. 9 (c) been used, the patterns of Fig. 10 would have been one-dimensional (horizontal) zone-plates instead of the equally spaced patterns shown, and any one of these horizontal line zone plates could then be used to reconstruct the echoing target. The signal-to-noise gain associated with the holographic (synthetic aperture) procedure would thereby be realized. Whereby the entire line of the zone plate, extending over a time period during which the target P of Fig. 9 (c) has traversed a sizable distance, would be utilized (coherently). If, in addition, a conical prism-lens is used, designed to focus, to the same focal point, the laser light diffracted by all of the horizontal line zone plates, a further signal-to-noise gain results.

A coherent variant of the above, again useful when power limitations exist in the transmitter components, is that of employing a multiplicity of coherent single-frequency signals (separated in frequency). In this case the original coherent oscillator signals would be used to generate (e.g., by synchronous demodulation), the multiple holographic zone plate interference patterns. The filtering process is thereby avoided, but the special conical prism-lens would still be required. The technique could be useful in geometries other than the bistatic case (where coherence length requirements are quite modest), including the standard (moving) synthetic aperture hologram systems (Fig. 9 (b)), and stationary systems operating against moving targets (8).

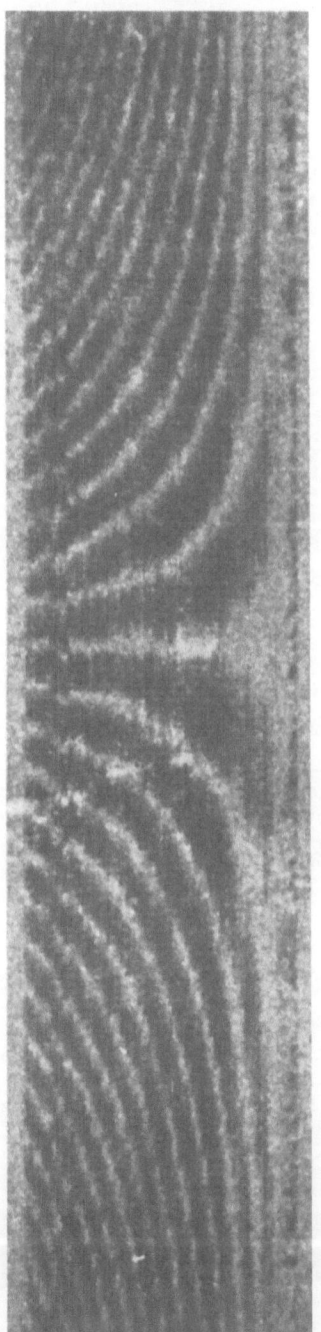

Fig. 11. When a noise source is made to interfere with itself and the combined signal analyzed with a narrow filter, the narrow band filter output behaves like a coherent signal, exhibiting for each frequency, constructive and destructive interference "fringes." (Frequency is plotted vertically, time horizontally).

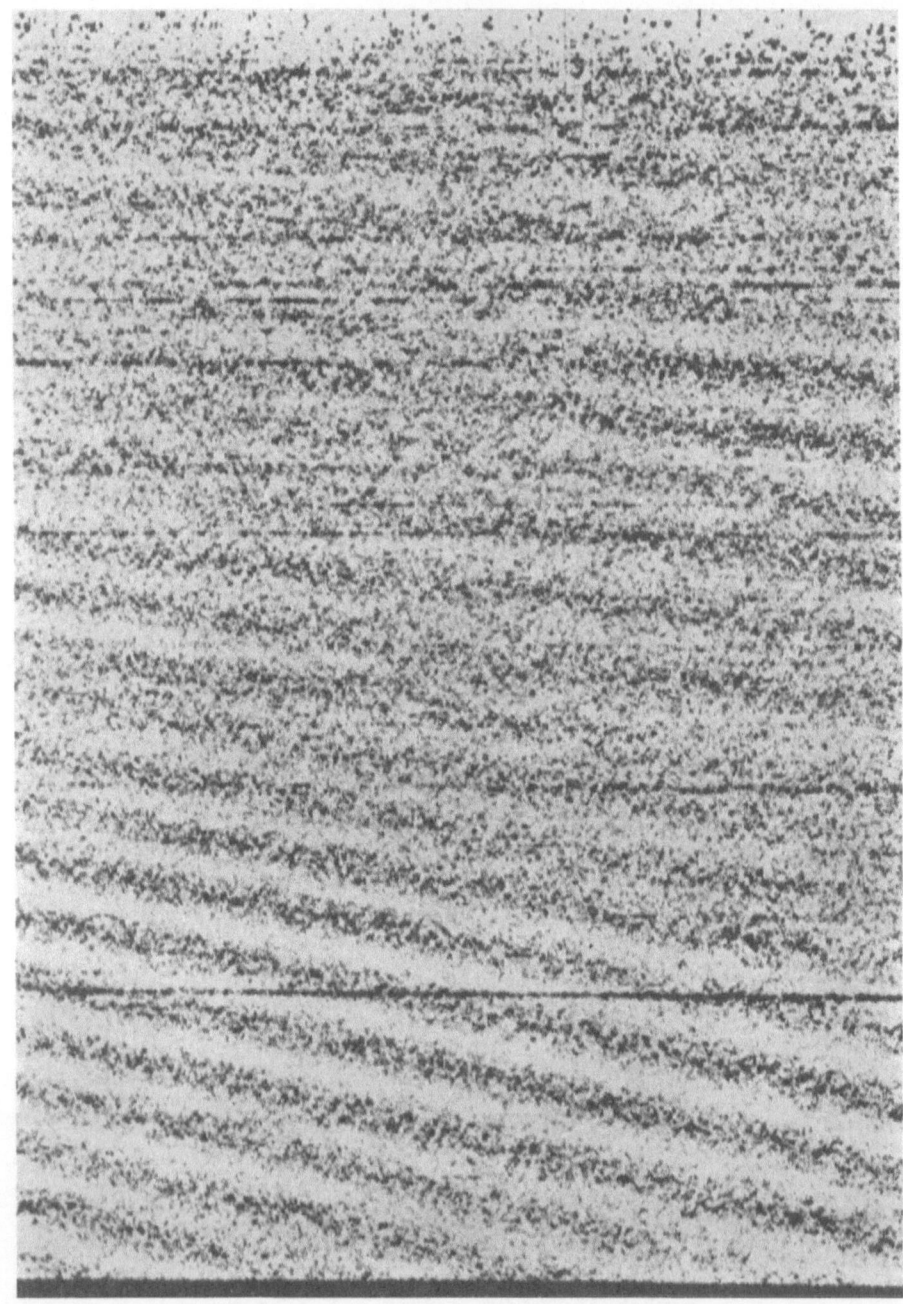

Fig. 12. An FFT spectrogram portraying an underwater sound interference pattern (from reference 17).

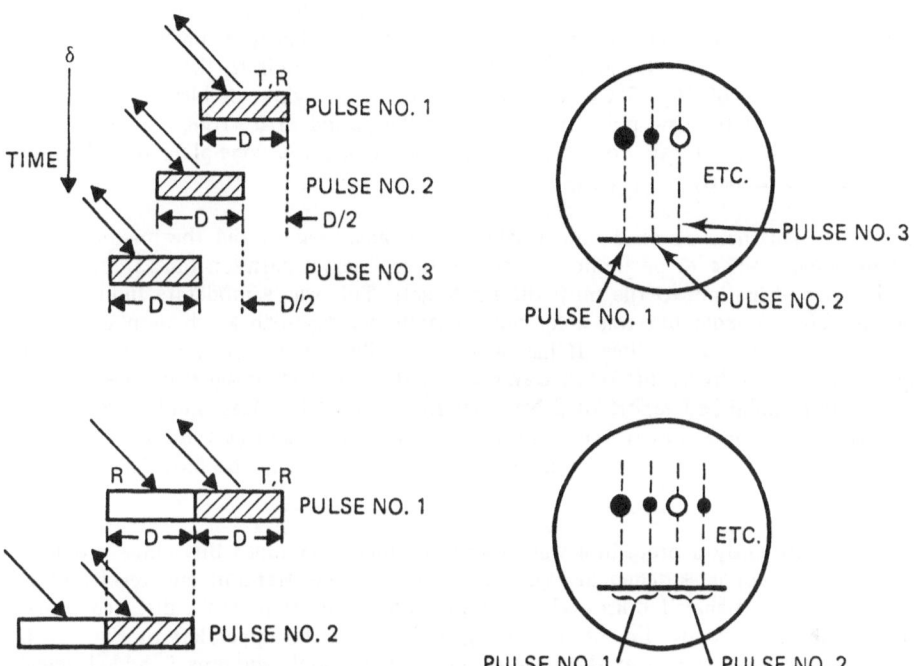

Fig. 13. In a standard synthetic aperture (hologram) sonar (top left) a new pulse must be transmitted each time the moving platform advances one-half the aperture dimension D. Top right indicates how a target generates, with successive pulse echoes, a horizontal one-dimensional zone plate on the cathode ray tube of Fig. 6. The heavy black dot corresponds to a constructive interference situation and the white circle to a destructive one. The lower sketches show how the pulse spacing (and maximum range) can be doubled through the addition of a receive-only transducer whose signal is fed to a second cathode ray tube beam moving up with the original one.

EXTENDING THE RANGE OF SYNTHETIC APERTURE SYSTEMS

One of the limiting factors in the synthetic aperture process results from the need to delineate accurately the finer fringes of the zone plates. The very close spacing of these finer fringes is readily evident at the right of the zone plate of Fig. 4. As the aircraft carrying the radar moves, the distance to the reflecting object changes, and a round-trip change of one half-wavelength causes the interference between the fixed reference wave and the varying eacho wave to change from a constructive interference case to a destructive one. The former corresponds to the black portions of the zone plate of Figs. 3 and 4 and the latter to the white or blank portions. It is obvious that if too few pulses return during the period of a "fringe" (adjacent white and dark areas), say, for example, only one pulse, the zone-plate record cannot be delineated properly.

This need to send out closely spaced pulses has caused the maximum useful (unambiguous) range to be limited, since the time interval between pulses corresponds to the round-trip time to the most distant targets. This was a moderate limitation on subsonic aircraft radar but was a far more serious one for high Mach number platforms, including aircraft and satellites. It has also been serious in the sonar case because of the much lower velocity of sound waves versus that of electromagnetic waves. Thus, in a recently published report of a National Academy of Sciences Summer Study, it was noted that for a ship travelling at a speed of 6 knots and carrying a 1 KHz, synthetic aperture, sonar, the maximum unambiguous range would be only 1.33 nautical miles.

Recently a procedure was described which overcomes this range problem (18). The concept is sketched in Fig. 13. The usual single transmit and receive case, with its need to transmit pulses whenever the antenna radar moves a distance equal to one half the aperture dimension D (19), is shown in the upper left of Fig. 12. In the new procedure, shown at the bottom left, a receive-only antenna is added ahead of the transmit-receive one. It is evident that the signal received by this second unit for pulse number 1 is the same as that received for pulse number 2 in the usual (upper left) case since for that case, the outward path is shorter and the receive path is longer. The pulse repetition rate can therefore be halved. If three receive-only antennas are used, the pulse repetition rate can be reduced by a factor of four (with a consequent increase in the maximum unambiguous range by that same factor), with seven receivers, by a factor of eight, etc. The cathode ray tube patterns (comparable to those of Fig. 2) for the two cases are shown at the right, for the second case, the tube must be equipped with two adjacent upward-moving beams, with the second one amplitude-modulated by the receive only transducer signal (properly combined with the reference wave, i.e. synchronously demodulated). It is seen that this recent development materially extend the usefulness of holographic (synthetic aperture) systems.

PARALLEL PROCESSING ASPECTS OF SYNTHETIC APERTURE SYSTEMS

It is of interest to compare the parallel aspects of the detecting, processing, and displaying functions of synthetic aperture (hologram) systems with the more usual serial procedures employed in present radars and sonars. In the detecting (accepting) function, the parallel nature is exemplified (1), in the multi-beam forming process (as against the serial beam-scanning technique), (2) in the parallel phase-curvature generating process for near-field points (versus the serial phase-adjusting, focussing procedure), (3) in the automatic parallel adjusting of aperture size so as to cause the metric resolution to be independent of range, and (4) in the providing of an angular resolution which is twice that of an ordinary, radar or sonar of equal aperture. In the processing function, the parallel nature is evident in the simultaneous generation of myriads of zone plates for points at all ranges and in the photographic process which records these zone plates. In the displaying function, the parallel process is manifest in the holographic reconstruction of the reflecting points through the illumination, with laser light, of the photographic record. These points were discussed in detail by the author at the 1973 NATO Advanced Study Institute on the Parallel Processing of Information, held in Capri, Italy June 17th to 30th under the direction of Professor E.R. Caianiello, Director of the Laboratory for Cybernetics at Naples, Italy. Papers presented there are to be published in book form (Plenum Press).

REFERENCES

(1)   L. J. Cutrona, E. N. Leith, L. J. Porcello, W. E. Vivian, "On the Application of Coherent Optical Processing Techniques to Synthetic Aperture Radar", Proceedings of the IEEE, Vol. 54, no. 8, pp. 1026-1031. (Aug. 1966).

(2)   W. E. Kock, "Side-Looking Radar, Holography, and Doppler-free Coherent Radar", Proceedings of the IEEE, Vol. 56, pp. 238-239, (Feb. 1968).

(3)   W. E. Kock, "Radar and Microwave Applications of Holography", Applications of Holography, pp. 323-356, (Plenum Press, New York, 1971).

(4)   J. J. Flaherty, K. R. Erikson, and Van Metre Lund, "Synthetic Aperture Ultrasonic Imaging Systems", U. S. Patent # 3, 548, 642, December 22, 1970 (filed March 2, 1967).

(5)   D. F. Pekau and R. Diehl, "Recording of One Dimensional Holograms as a function of Object Range", presented at the International Symposium on Applications of Holography, Besançon, France, July 6-11, 1970.

(6)   W. E. Kock, "Synthetic End-fire Hologram Radar", Proceedings of the IEEE, (Letter), vol. 58, November 1970, pp. 1858-1859.

(7)   W. E. Kock, "Holographic Techniques in Continuous-wave Bistatic Radars", Proceedings of the IEEE, (Letter), Vol. 58, November 1970, pp. 1863-1864.

(8)   W. E. Kock, "A Holographic (Synthetic Aperture) Method for Increasing the Gain of Ground-to-air-Radars", Proceedings of the IEEE, (Letter), vol. 59, no. 3, March, 1970, pp. 426-427.

(9)   W. E. Kock, Letter, to appear shortly in the Proceedings of the IEEE.

(10)   H. Ogura and K. Iizuka, "Hologram Matrix and Its Application to a Novel Radar", Proceedings of the IEEE, (Letter), vol. 61, pp. 1040-1041, July, 1973.

(11)   W. E. Kock, "Lasers and Holography, An Introduction to Coherent Optics", Doubleday, 1969.

(12)   W. E. Kock, "New Forms of Radar and Ultrasonic Imaging" in "Ultrasonic Imaging and Holography", editors G. W. Stroke and W. E. Kock, Plenum Press, 1974.

(13)   W. E. Kock, "Acoustical Holography", in Vol. 10 of "Physical Acoustics", Editors W. P. Mason and R. N. Thurston, Academic Press, to appear in 1973.

(14)   W. E. Kock, "Bistatic Microwave or Acoustic Holography Using Incoherent Illumination", Proceedings of the IEEE, (Letter), to appear in October issue.

(15)   Winston E. Kock, "Seeing Sound", New York: Wiley-Interscience, 1971.

(16)   Such FFT methods are described in W. E. Kock, "Radar, Sonar, and Holography". New York: Academic Press, to appear in 1973.

(17)   G. D. Bergland, "A guided tour of the fast Fourier transform", IEEE Spectrum, vol. 6, pp. 41-52, July 1969.

(18)   W. E. Kock, "A Method for Extending the Maximum Range of Synthetic Aperture Radar", to appear in the Proceedings of the IEEE, (Letter).

(19)   W. M. Brown and L. J. Porcello, "An Introduction to Synthetic Aperture Radar", IEEE Spectrum, September 1969, pp. 52-62.

P.S.   This paper was not presented at the conference.

HOLOGRAPHY BY SHADOW CASTING

H. J. Caulfield

Block Engineering, Inc.

Cambridge, Massachusetts (U.S.A.)

1.   INTRODUCTION

"Coded aperture imaging" is a recently agreed-upon name (1) for a two-step imaging process conceived in the early 1960's by Mertz and Young (2, 3, 4, 5) at Block Engineering. Their basic idea was to let each point in the object encode its three-dimensional location by casting a shadow of a Fresnel zone plate onto a photographic plate. Thus an object consisting of a collection of points would be encoded as a collection of zone plates. Lateral displacement of the object points leads to lateral displacement of the shadow-cast zone plates. Vertical displacement of the object points leads to scale changes in the shadow-cast zone plates. Given this encoded image, it is easy to arrive at an image. Illumination with coherent light converts each zone plate into a point. The lateral position of the decoded point is determined by the lateral position of the corresponding zone plate. The vertical position of each decoded point is determined by the scale of the corresponding zone plate. Thus the three-dimensional image encoded by shadow casting can be decoded by illumination with coherent light.

Encoded aperture imaging was intended to offer certain advantages over other imaging techniques for wavelengths shorter than those of visible light, e.g. for γ-rays or x-rays. The intended advantage is in efficient utilization of the available radiation. The open area of a pinhole aperture and the open area of a Fresnel zone plate aperture yielding the same resolution are quite different. In both cases the resolution is limited by the smallest dimension of the aperture, i.e., the pinhole diameter or the smallest zone width. We will show later that an N-zone Fresnel zone plate has ($\sim N^2$) times the open area of a pinhole giving equivalent resolution.

There were a number of problems with encoded aperture imaging as it was originally practiced, but one problem was especially annoying. The technique failed to produce images of extended objects. For this reason applications appeared to be limited to x-ray and γ-ray astronomy where the typical "object" is a small collection of discrete "points."

Although considerable effort was applied the situation remained unchanged for a decade.

Then in 1972 Barrett and his coworkers at Raytheon announced a breakthrough. They had made excellent images with extended objects (6, 7, 8).

As often happens, that breakthrough redirected interest to a neglected field and stimulated new investigations. In 1973 we are confronted with an array of encoded aperture imaging techniques. These include the method due to Barrett et al, the original Mertz and Young method (now found to work well in certain cases), and many others.

In this chapter, I hope to do two things. First, I want to review the state of the art so readers will know what can be done with presently-available methods. Second, I want to survey some of the problems and possibilities so readers may join in research in this area.

Two papers are recommended to the reader for material not covered in detail in this chapter. The first is a theoretical paper by Barrett and Horrigan (9). This paper is important in several respects. It contains derivations of most of the important relationships for coded aperture imaging. Also, by agreement among many workers in the field (1), it will be used as the source of notation and terminology for subsequent papers in this area. The second is a review by Caulfield and Williams (10) which gives more details on the history of coded aperture imaging and on the relationships between coded aperture imaging and holography.

2.   FRESNEL ZONE PLATES

In its original forms and in most of its current ones, coded aperture imaging uses Fresnel zone plates as apertures. Therefore, a brief digression on the properties of Fresnel zone plates (FZP's) is a useful preliminary exercise.

An ideal lens can be considered to be a device which multiplies the incident wavefront by a quadratic phase factor. That is the ideal lens has an effective transmission of

$$\ell\,(r) = \exp\,(ikr^2/2f), \qquad\qquad\qquad\qquad 1$$

where r is the radial coordinate, k is the wavenumber ($2\pi/\lambda$), and f is the focal length of the lens. Of course, we can write

$$\ell_I\,(r) = \cos\,(kr^2/2f) + i\,\sin\,(kr^2/2f) \qquad\qquad\qquad 2$$

Now it is possible to retain the lens-like behavior (while introducing an ambiguity in the sign of f) by using the cosine term alone. It would be desirable to record

Fig. 1. A Fresnel zone pattern.

sin $(kr^2/2f)$ directly but we know of no convenient way to do so. By adding a coherent reference beam we can record a hologram of transmission

$$\ell_H(r) = \frac{1}{2} + \frac{1}{2} \sin (kr^2/2f) \qquad\qquad 3$$

A further simplification is to simply replace the negative parts of the cosine function by zero to obtain a recordable positive function. One more drastic simplification leads to the FZP. We replace all of the positive values of sin $(kr^2/2f)$ by 1. So the transmission of a FZP is

$$t_{FZP}(r) = \begin{cases} 1 \ \text{if} \ \sin \ (kr^2/2f) > 0 \\ \\ 0 \ \text{if} \ \sin \ (kr^2/2f) < 0 \end{cases} \qquad\qquad 4$$

Figure 1 shows a FZP pattern. Most of its properties are quite familiar. It consists of zones bounded by concentric circles of radii

$$r_n = r_1 \sqrt{n} \qquad\qquad 5$$

where r_1 is the radius of the central disc. As described above, the central disc would be clear ($t_{FZP} = 1$). The complementary pattern (dark central disc) is also called a FZP. It corresponds to the imaginary part [sin $(kr^2/2f)$] of the transmission of an ideal lens and hence leads to an image π out of phase with the cosine FZP image.

It is clear that the area of the nth zone is

$$A_n = \pi r_n^2 - \pi r_{n-1}^2 = \pi r_1^2 \ [(n) - (n-1)] = \pi r_1^2 = A_1 .$$

That is, the area of each zone is equal to the area of the central disc.

There are numerous generalizations of FZP's. I will mention two in passing. First, we can change the quantization level. Thus we get

$$t_{FZP,a} = \begin{cases} 1 \ \text{if} \ \sin \ (kr^2/2f) > a \\ \\ 0 \ \text{if} \ \sin \ (kr^2/2f) < a \end{cases}$$

Of course, a can be any number between 1 and -1. Second we can change the Θ dependence but not the r dependence. Thus instead of exp $(ikr^2/2f)$ we could use exp $(iN\Theta/2\pi)$ exp $(ikr^2/2f)$. This destroys the complete rotational symmetry of the FZP and changes some of its imaging properties.

3.   SHADOW CASTING

The shadow casting process can be described exactly if two circumstances hold. First, the geometry must exclude significant diffraction effects over the shadow casting distance. A useful criterion is

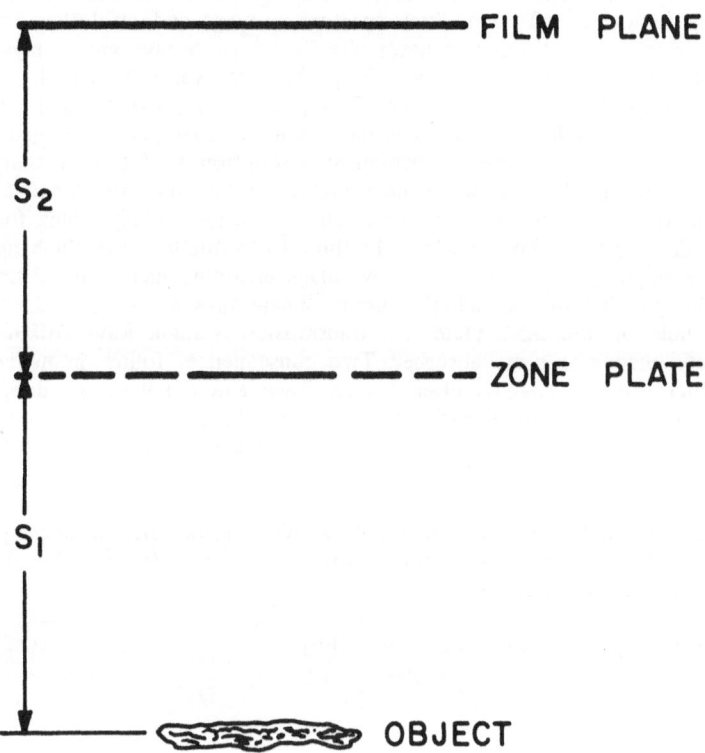

Fig. 2. The geometry for shadow casting.

$$d_s < a^2/\lambda$$

where d_s is the shadow casting distance, a is the radius of the smallest opening in the aperture or dimension of the object, and λ is the wavelength. Let us make a typical calculation. Suppose $\lambda = 0.5 \times 10^{-4}$ cm (green light) and a = 0.1 cm. Then we require $d_s < 200$ cm. The importance of small λ is now evident. For γ rays and x rays almost any objects, aperture, and distances can be used because λ may be 10^{-7} or 10^{-8} cm. With $\lambda = 10^{-7}$ cm we can operate at distances of 10 cm with spacings as low as 10^{-3} cm. Such aperture spacings and resolution are impractical, so the diffraction criterion is never checked when γ rays and x rays are being used. At the other extreme, waves cast sharp shadows over a 10 cm distance only for aperture spacings of 10 cm or more. This probably represents about the greatest wavelength at which coded aperture imaging might be attemped. The second condition for the applicability of the mathematical description to follow is that the shadow casting be linear. That is, we assume that all points on any plane parallel to the coded aperture cast the shape pattern. This condition usually holds for visible light where the aperture may be relatively thin. Unfortunately the thick aperture plates usually required for x-ray and γ-ray image encoding have transmission which vary with the position of the radiating point. Where rays are incident in the direction of the holes in the thick plate, the transmission is high. Rays striking the same hole obliquely encounter more absorber. Two consequences follow immediately: (1) The linear mathematical analysis often breaks down and (2) we must design the aperture to minimize this nonlinearity since a linear image decoding will be assumed. The minimum step required is to keep the holes at least as large as the thickness of the plate.

The geometry is shown in Figure 2. We consider first a single point object located at (x, y) and an aperture with transmission g(x', y'). Clearly the shadow due to that point is

$$S_o(x'', y'') = g\ (ax'' - bx,\ ay'' - by)$$

where

$$a = S_1/(S_1 + S_2)$$

and

$$b = S_2/(S_1 + S_2).$$

The minus sign indicates that shifting the point to the left shifts the shadow to the right, etc. That is, the rays pivot at the aperture. Now we make the linearity assumption. That is, that the shadow is the sum of the individual $S_0(x'', y'')$ shadows from each point in the object. Thus the shadow due to an object given by f(x, y)

$$h\ (x'', y'') = \iint\limits_{\text{object}} f\ (x, y)\ g\ (ax'' - bx,\ ay'' - by)\ dxdy.$$

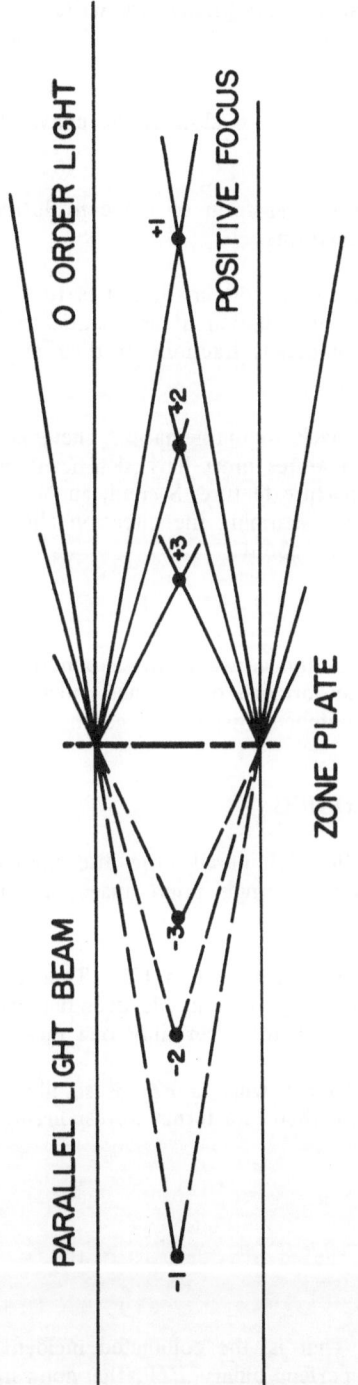

Fig. 3. Optical effect of a Fresnel zone plate.

Neglecting such niceties as scale factors and primes, we write

$$h(x, y) = f(x, y)* g(x, y)$$

where * indicates convolution. That is the shadow is the convolution of the object function with the aperture function.

We know that the Fourier transform of the convolution of two functions is the product of their Fourier transforms.

Translating this last statement from mathematics to English, we discover that the shadow has good depth of modulation at any spatial frequency if and only if $f(bx, by)$ and $g(ax', ay')$ have significant fractions of their energy at the same spatial frequency.

We must now translate back to mathematics. There are two requirements for sharp shadowing. First, object features must have significant components along the direction of some prominent aperture feature. Second, an object feature of size ΔX will contribute a strong shadow (assuming the direction alignment just prescribed) if there is an aperture feature of size about

$$\Delta X' = b\Delta X$$

Objects, apertures and geometries jointly satisfying those conditions are said to be cases of "matched imaging." The importance of matched imaging can not be over-emphasized. Without approaching matched imaging we get no usable shadow.

4.   OPTICAL IMAGE DECODING

Our discussion of decoding will parallel the discussion of image encoding. That is, we will discuss the decoding of single point images and then use linearity to describe decoding of complex objects.

The coded image of a point source if a FZP. Thus any linear method which converts the FZP to a point image is a usable decoding method. Illumination of the FZP with a coherent beam of light is certainly one means.

Figure 3 shows what happens when a FZP is so illuminated. Besides the strong undiffracted (zero-order) term, there are terms corresponding to focal lengths

$$f_m = \pm\, r_1^2/\lambda m$$

where

$$m = 0, 1, 2, 3, \ldots$$

Of course, for $m = 0$, $f_o = \pm\infty$. That is, the collimated incident light remains collimated. Actually if we have a perfect binary FZP, the non-zero even m's do

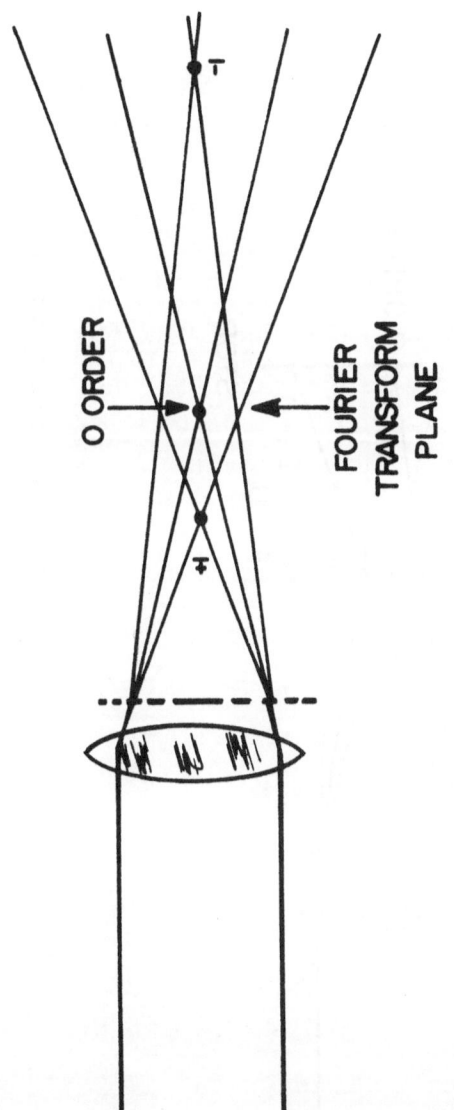

Fig. 4. Image decoding for on-axis Fresnel zone plate encoding.

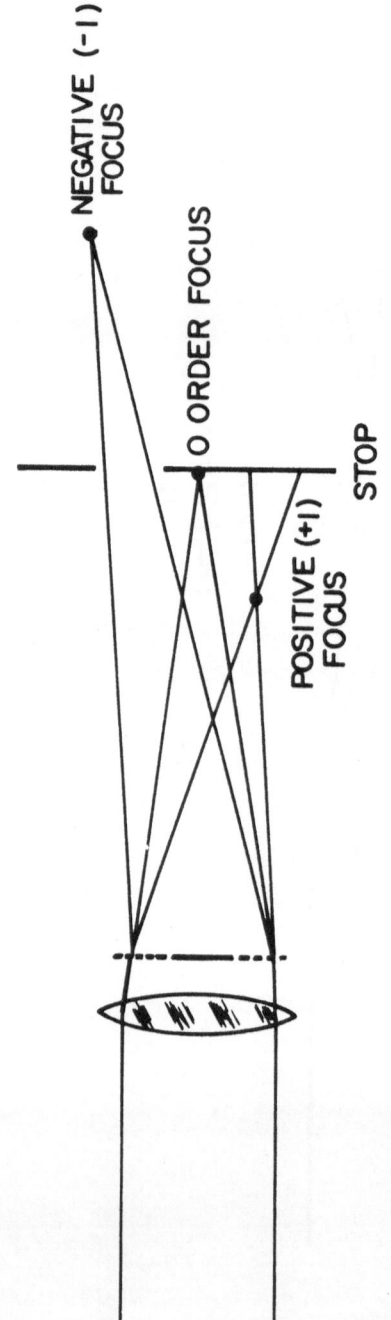

Fig. 5. Image decoding for off-axis Fresnel zone plate encoding.

not lead to actual images because the even-ordered terms in the Fourier series of a centered square wave are zero. In most real cases all orders are present. Usually r_1 is large in h(x, y), so a demagnified shadow is used for image decoding. We have written

$$h(x, y) = f(x, y) * g(x, y).$$

The linear operation represented by illumination with coherent light can be called I [·]. We know

$$I [g (x, y)] = p (x, y),$$

where p(x, y) is the point spread function of the aperture used as a focusing device. Then

$$I [h (x, y)] = I [f (x, y) * g (x, y)]$$

$$= f (x, y) * p (x, y)$$

$$\cong f (x, y).$$

That is, coherent light illumination leads directly to image formation.

The width of p(x, y) by the Rayleigh criterion is

$$d = \beta \lambda f_1 / D_1$$

where β is a unitless numerical factor and D is the diameter of the zone plate. As N becomes large ($>$10), we can approximate β quite accurately with the diffraction-limited value of 1.22.

One of the primary image decoding problems is the existence of all of the f_m's other than the particular one used for image processing. Two approaches have been taken. First, Mertz and Young (2, 3, 4, 5) blocked the unwanted beams by spatial filters. Second, Tipton et al. (11) caused destructive interference between the uniform zero-order beam and the first-order image. There follows a brief review of these approaches.

Spatial filtering of unwanted orders can occur in either of two ways. First, if an ordinary FZP is used, the arrangement shown in Figure 4 suffices. Unfortunately this blocks much of the image information also, so this method has fallen into disfavor. Second, if an off-axis section of a FZP is used, the arrangement in Figure 5 is ideal. The trouble is that most objects do not have the spatial frequency content to cast sharp shadows of an off-axis FZP. Therefore, this method too fell into disfavor. The breakthrough of Barrett et al was to introduce a half-tone screen between the object and the coding aperture to impress the proper spatial frequency onto the object. Thus they revived interest in the off-axis FZP case.

Destructive interference between the first-order beam and the zero-order

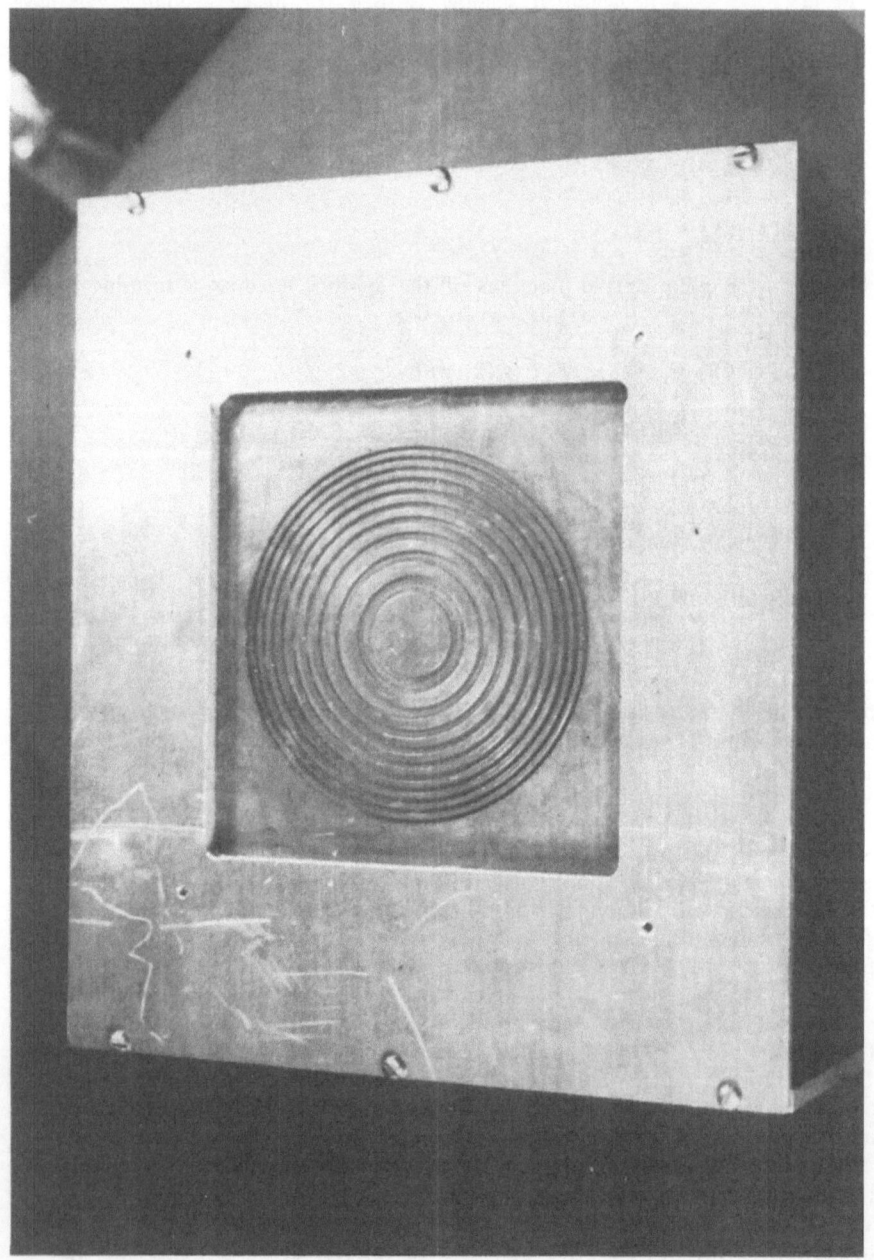

Fig. 6. Photograph of the zone plate used to encode γ-ray images.

beam occurs if the reduced shadow FZP due to each object point is dark centered. Then simply illuminating the reduced shadow image by a collimated beam of coherent light results in a negative, high-contrast image a distance f_1 away.

5.   TYPICAL RESULTS

In this section we describe some typical experiments and cite the results.

All images were encoded using the zone plate shown in Figure 6. Radiation in the form of 140 keV γ-rays from a source of 99m Tc was used in all cases. The objects were hollowed-out plexiglas phantoms filled with the isotope.

In each case the shadow was recorded as a negative on DuPont Cronex II X-ray film using a DuPont "Lightning Plus" intensifying screen. The encoded record was reduced optically by about 30 times, and the final record was the negative recording of the reduced image. Having passed through two successive negative recordings, the encoded record was positive on the demagnified record. Thus the dark-centered zone plate leads to a dark-centered demagnified record.

To retrieve the image, the unbleached demagnified record was inserted in a collimated beam drived from a He-Ne laser. The image formed at the appropriate distance and interferred with the coherent background to produce a sharp negative reproduction of the object.

Most of the advantages of γ-ray coded aperture imaging seem to be autoradiography, where one attempts to visualize specific internal organs (e.g., brain, liver, thyroid) by the uptake of a particular radioisotope. The Picker (NI-94198) thyroid phantom is the standard object agreed upon for image quality comparisons (1). We used S_1 = 6.5 cm and S_2 = 18.0 cm to encode its image. A photograph of the decoded image is shown in Figure 7. All features of the phantom are easily seen. These include three "cold" nodules represented by plastic cylinders within the phantom, a "hot" region at the lower end of the right lobe, and an overall more active left lobe. A small screw at the lower region of the right lobe causes the apparent shortening of that lobe.

Two further experiments demonstrate the insensitivity of this technique to object geometry. First, two objects of comparable spatial frequency content by different distances from the recording medium were imaged. For the "S" objects S_2 = 15.3 cm and for the "+" object S_2 = 20.3 cm. S_1 = 5.7 cm for both objects. The stroke width on each object was 0.32 cm. Figure 8 shows the reconstructed wavefront pattern in two planes. Both objects are reproduced clearly. Second, an object with a wide range of spatial frequencies was imaged using a fixed taking geometry (S_1 = 13 cm and S_2 = 20 cm). The image is shown in Figure 9.

6.   CURRENT STATE OF THE ART

At present the available films, intensifiers, etc. restrict the sensitivity of coded aperture imaging to values of 3 to 10 times less than those of the best avail-

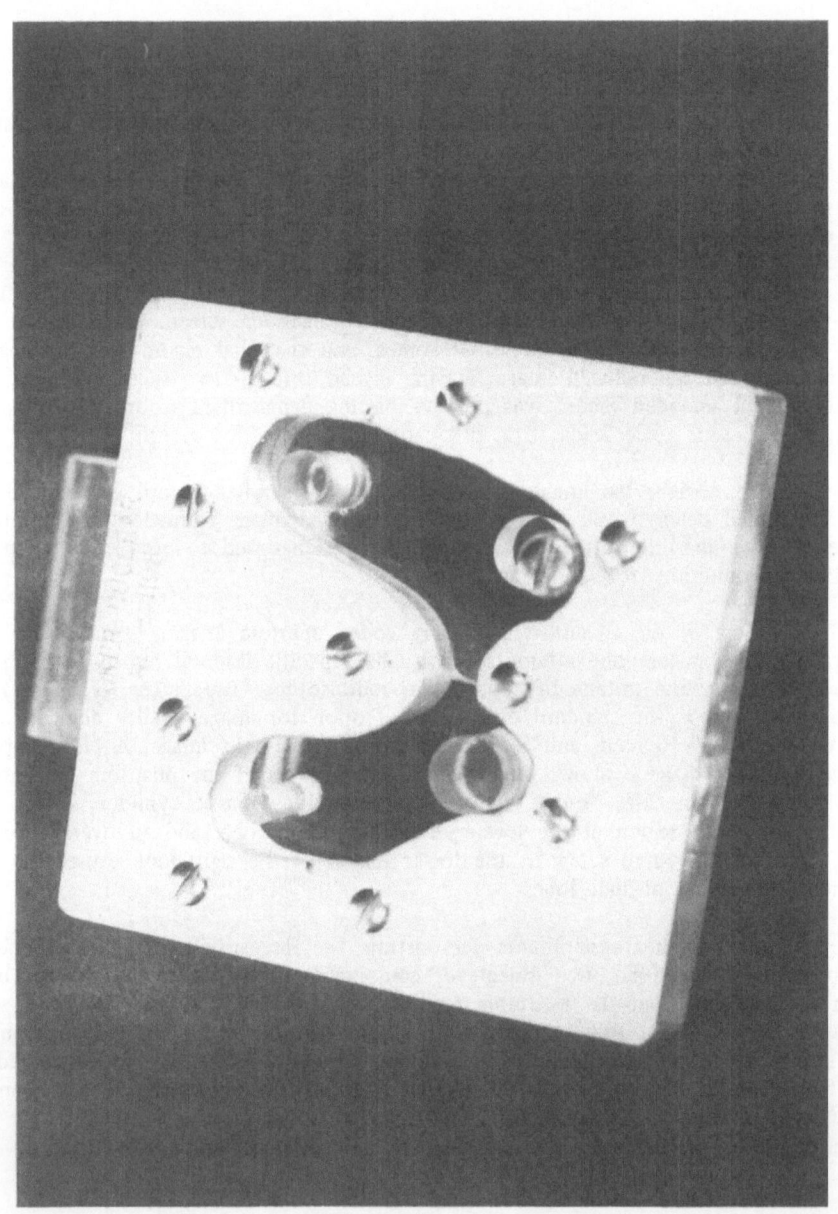

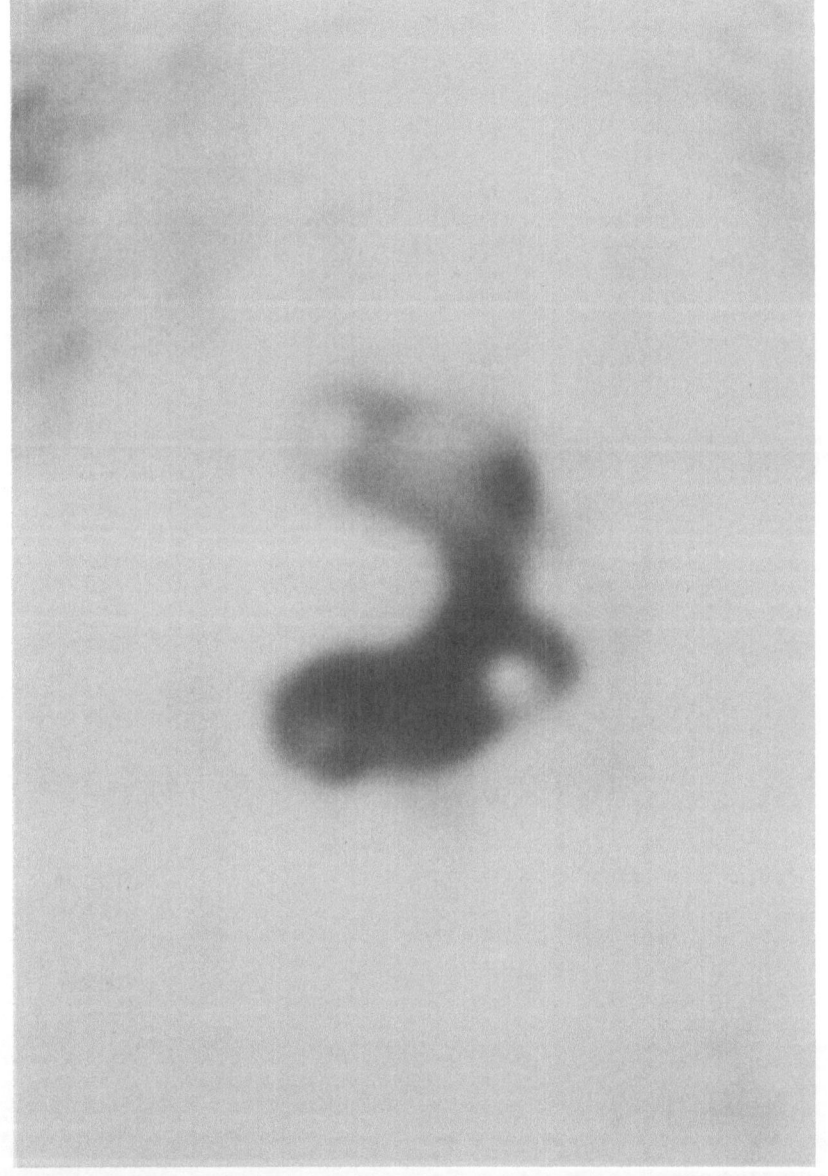

B

Fig. 7. Photograph of a thyroid phantom (A) and its image produced by encoded aperture methods (B).

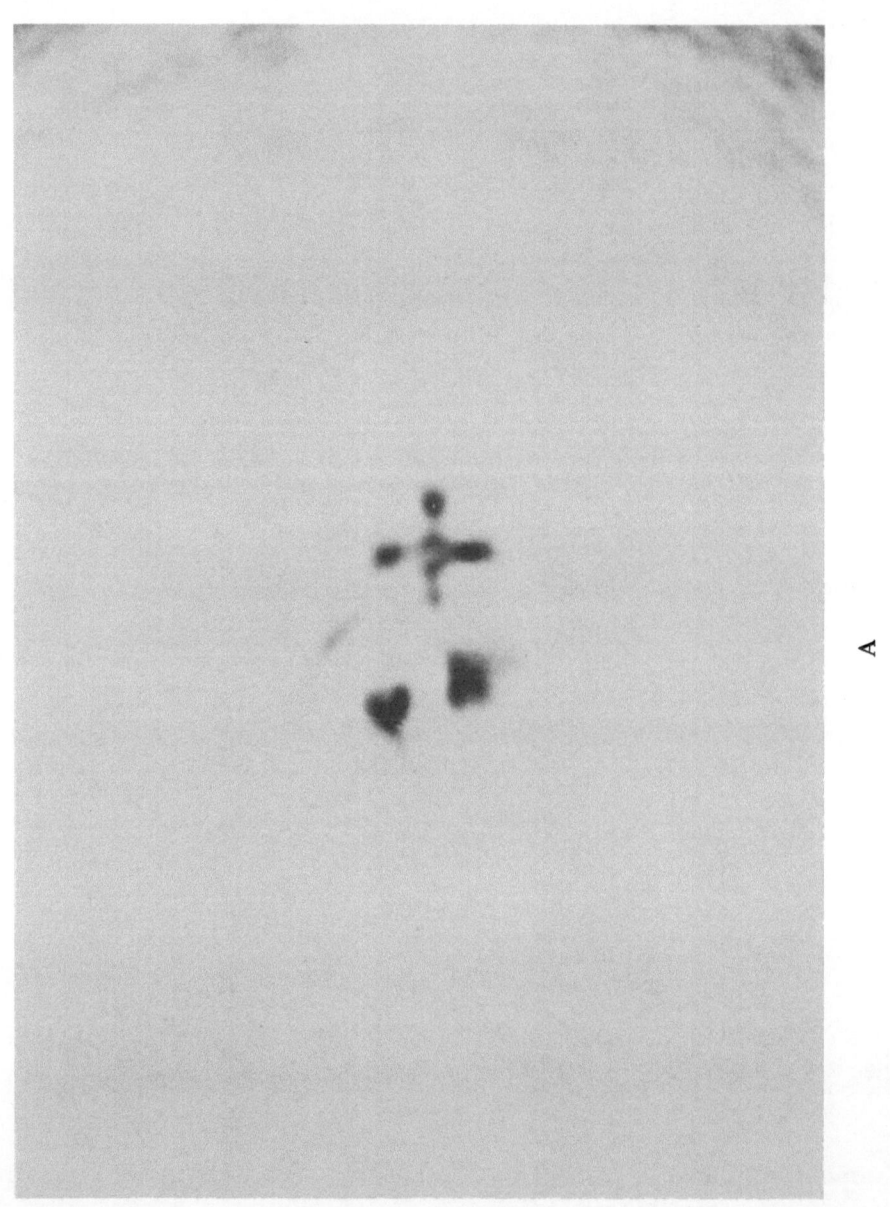

A

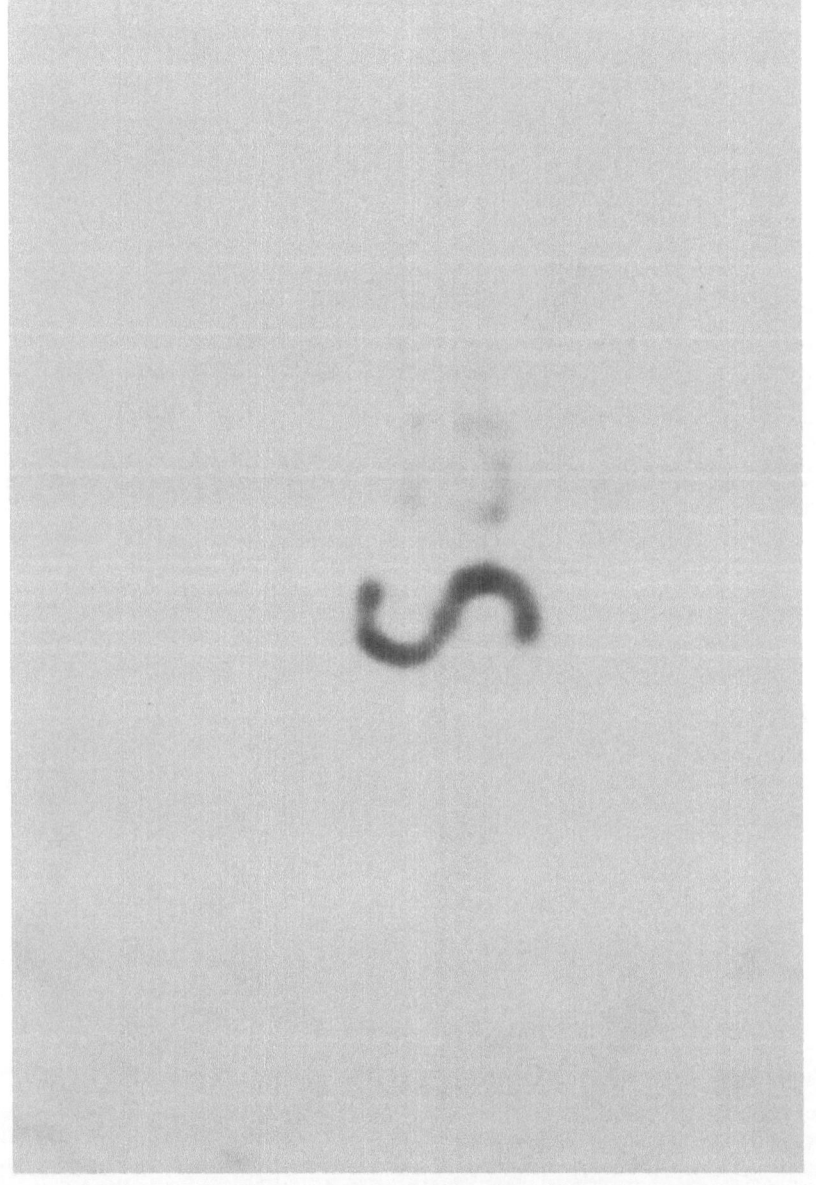

B

Fig. 8. Two objects separated in depth by 5 cm led to the spatially separated "+" and "S" images shown here. The out-of-focus image of the other object is visible in each case.

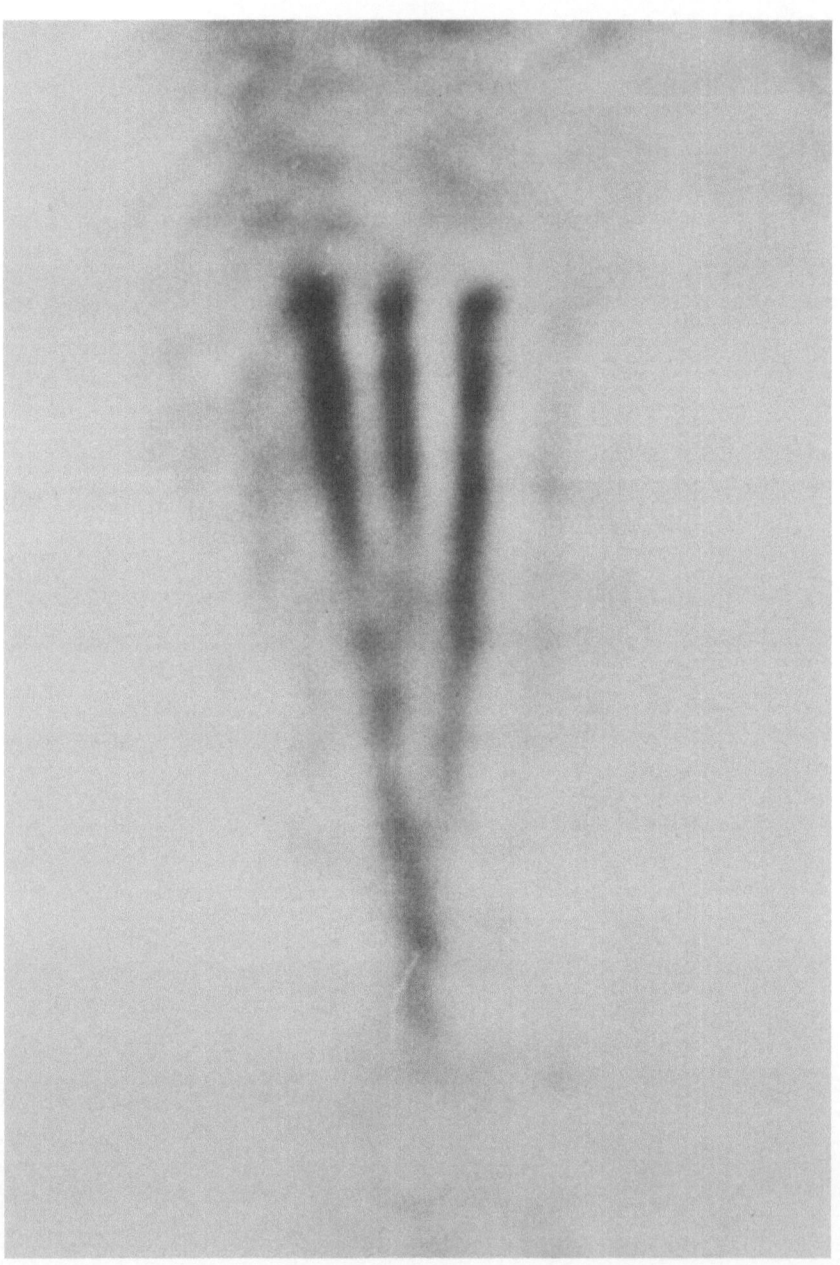

Fig. 9. Image of an object with a wide range of spatial frequencies.

able Anger cameras. It can be shown that no signal-to-noise advantage over a pinhole camera can be obtained if the recording is shot noise limited. There are several advantages though. First, the equipment for coded aperture imaging is very simple. The open area of a FZP is $8N^2/\beta^2$ larger than the open area of a pinhole giving equal resolution. The factor is somewhat smaller for off-axis FZP's. This increased throughput allows us to use relatively simple detection means like film. Thus coded aperture imaging offers a cost advantage over any equally sensitive imaging technique. Second, the image is three-dimensional. Because of the small size of the reduced shadow and the small number of zones, the three-dimensional image cannot be viewed directly as with coherent optical holograms. Rather we must view one depth plane at a time.

7.   CONCLUSION

I have attempted a broad introduction to a field of great current interest. There are many practical problems which I have not mentioned. For each problem there are several proposed solutions. Activity in this field is now quite widespread, so many new developments can be expected within the next several years.

REFERENCES

(1)   The text of the agreements reached by most American workers in coded aperture imaging will be published soon in Applied Optics. Preprints are also available from the author.

(2)   L. Mertz. J. Opt. Soc. Am., advertisements, Feb (1960) and May (1961).

(3)   L. Mertz, Transformations in Optics (John Wiley, New York, 1965).

(4)   N. O. Young, Sky and Telescope 25, 8 (1963).

(5)   L. Mertz and N. O. Young in Proc. Int'l. Conf. on Optical Instruments, (Chapman and Hall, London, 1961).

(6)   H. H. Barrett, D. T. Wilson and G. D. DeMeester, Optics Comm. 5, 398 (1972).

(7)   H. H. Barrett, K. Garewal, and D. T. Wilson, Radiology 104, 429 (1972).

(8)   H. H. Barrett, D. T. Wilson, and G. D, DeMeester, Opt. Eng. 12, 7 (1973).

(9)   H. H. Barrett and F. A. Horrigan, Appl. Opt. 12, (1973).

(10)   H. J. Caulfield and A. D. Williams, Opt. Eng. 12, 3 (1973).

(11)   M. D. Tipton, J. E. Dowdey, and H. J. Caulfield, Opt. Eng. 12, (1973).

LIQUID CRYSTAL MATRIX DISPLAYS

C. H. Gooch

Ministry of Defence, Service Electronics Research Laboratory

Baldock, Hertfordshire (U. K.)

1.   INTRODUCTION (1)

Electronic techniques of displaying information are becoming widely used. The applications of these displays range from the display of simple numeric information - as in a watch or digital voltmeter - to the display of radar or television information. Most of these displays emit radiation and thus consume considerable amounts of power when they are bright enough to be seen in high-ambient lighting conditions. The alternative display is one that does not compete with incident radiation but modulates it - just as a printed page or blackboard writing does. A display of this form will be visible in the brightest lighting conditions but need not consume large amounts of power. They are known as "passive" or "subtractive" displays and it is in this area that liquid crystals show considerable promise.

Many displays are formed from an array of elements, a selection of which is activated or "selected" to form a pattern that conveys the required information. Two simple examples of this are the numeral display consisting of seven bars in the form of a figure 8, and the more complex alphanumeric display which consists of 35 elements arranged in a 7 x 5 matrix.

The 35 element display can make an adequate representation of the 26 alphabet characters as well as the numerals 0-9 and a range of conventional signs. The simplest way of operating such a display is to take a separate lead to each element and apply a suitable signal to activate the required display elements. This is a viable approach with a 7-bar numeral but with the 35-element character the number of leads becomes inconveniently large. For bigger displays the number of

---

(1) Crown Copyright.

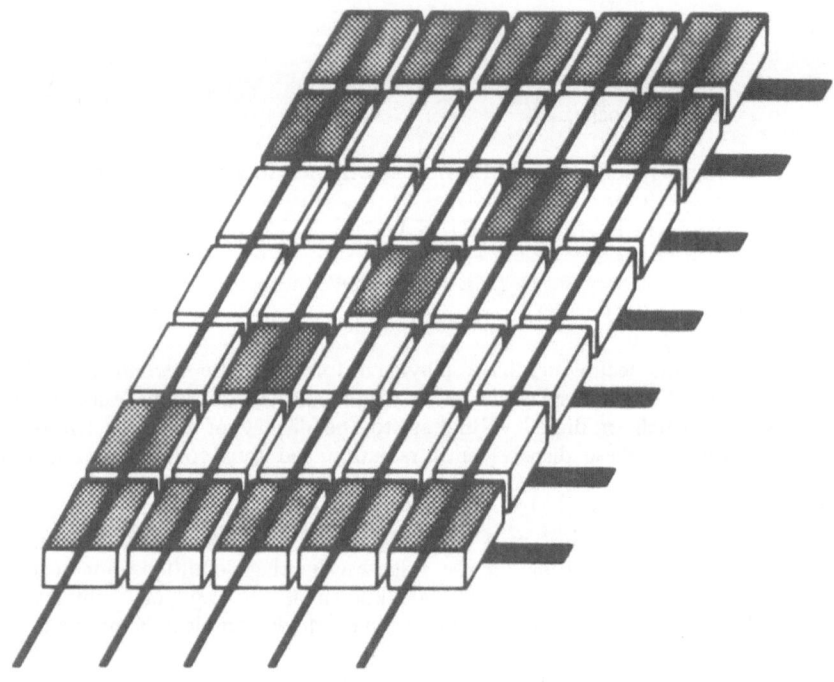

Fig. 1. The matrix display.

leads required - on a one per element basis - is a severe limitation on the feasibility of the system.

The alternative is to use a matrix or coordinate scheme to select or address the display elements as shown in Figure 1. Here the elements are connected by 7 row and 5 column leads, thus reducing the number of leads, in this case, from 35 to 12. For larger displays, for example those consisting of multiple characters, the reduction in the number of leads is even more significant.

The next section will discuss the basic requests of a matrix connected display and then, in sections 3 and 4, we will consider how these requirements can best be achieved using the electrooptic phenomena shown by liquid crystals.

2.   MATRIX DISPLAY PRINCIPLES

The use of matrix techniques to connect the elements of a display makes a considerable simplification to the display in some respects but also introduces a number of complications. These will be discussed in the following paragraphs and in this discussion it will be convenient to designate the rows and columns of the display a, b, c.... and p, q, r.... respectively.

2.1.   Selection Techniques

In a matrix display an element is selected by applying a suitable signal between one of the row and one of the column leads. In the familiar case of L. E. D. displays, in which the elements have a diode characteristic, this simple scheme introduces no problems because current can flow in only one path between a row and a column. That is the current must flow through the selected element since all other parallel current paths contain a diode element which is reverse biased and thus presents a high impedance to the current flow.

If the display elements have a linear resistive or capacitive characteristic we must take into account a number of alternative current paths which will cause elements other than the selected one to be partially activated and thus degrade the display. In considering this problem there are two methods of addressing the display that must be considered.

a.   Half-selection
The most common method of addressing a matrix display is that known as the "half-select" method, shown in Figure 2a. Here the voltage required to activate the display element is applied between the appropriate row and column while all other rows and columns are held at the mid-point potential, which is usually arranged to be zero or ground potential as shown.

In this situation the full voltage required to activate an element appears across the selected element but half that voltage also appears across the elements which are in the same row or column as the selected element. If the display phenomenon that is being used has a sharp threshold such that it is fully "on" for a

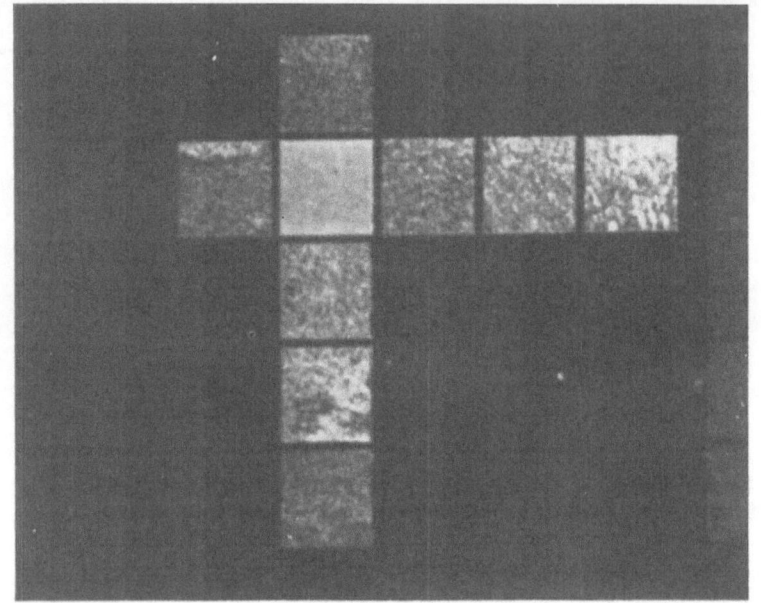

b

a

Fig. 2. Selection technique. (a) Half-selection; (b) Third-selection.

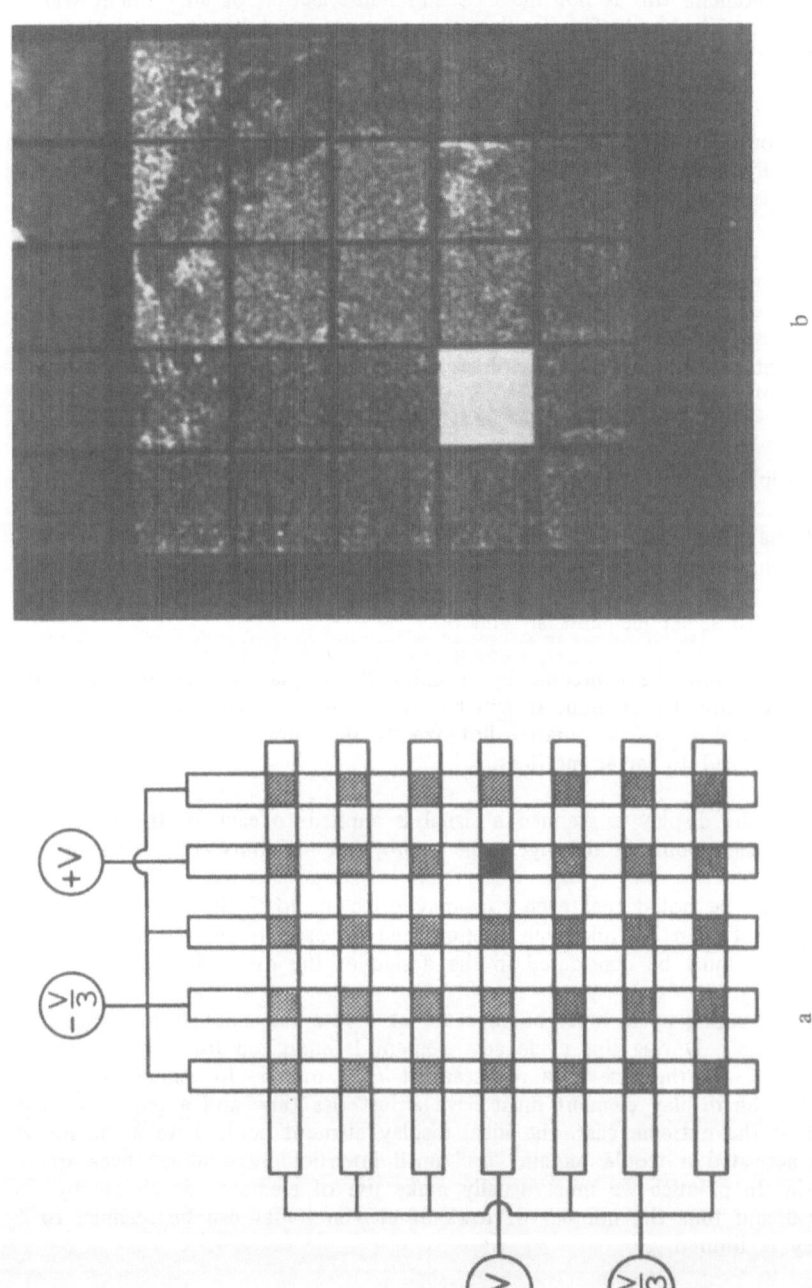

Fig. 3. Selection techniques applied to a liquid crystal display (dynamic scattering). (a) Half-selection; (b) Third-selection.

voltage 2V, but "off" for a voltage of V, this will cause no problem, but with most liquid-crystal phenomena this is not the case and half-selection of an element will cause a cross to be displayed. This is illustrated, for a typical liquid-crystal electro-optic effect, in Figure 2b.

b.   Third-selection

The problem of half-selection, which was described in the previous section can be reduced by using the addressing technique known as third-selection (Figure 3a). Here the unselected rows and columns are not grounded but connected to potentials ±V/3.

As a result of this a voltage of one-third the full drive voltage is applied to every unselected element in the display. The discrimination between selected and unselected elements is thus improved, as shown in Figure 3b and the feature that all unselected elements have the same voltage applied to them can be an advantage in some situations.

2.2   Multiplexing or Scanning

We have seen how one element in a display can be selected but in any practical situation a number of elements must be selected to form the display. This introduces a complication since, for example, the selection of two matrix elements ap and bq will also select elements aq and bp.

This can only be overcome by scanning the display so that it is built up row by row or column by column. It will be convenient to discuss the first of these alternatives as applied to a 7 x 5 matrix but exactly the same considerations apply to either situation and to larger matrices.

When the display is scanned a signal is applied to each of the rows in sequence and at each point in this cycle the appropriate columns are energized.

It must be noted that each row may drive up to 5 elements in parallel, for a duty cycle of 1 in 7, while each column lead drives only one element at a time. These points must be considered in the design of the drive circuits.

The scanning cycle must be repeated at a rate fast enough to avoid the appearance of flicker. During this cycle any element is energized for only a small duty cycle - 1 in 7 for the case of a row-scanned 7 x 5 matrix. In general terms this means that the display element must have a fast rise time and a relatively slow decay. Taken to the extreme case, the ideal display element could have a "memory" so that once activated it would remain "on" until intentionally switched back to the "off" state. In practice we must usually make use of elements which are by no means ideal, and thus the number of rows of elements that can be scanned to form a display is limited.

If the frame time of the display is T and the number of lines to be scanned is n, the time available to address an element (once per frame) is T/n.

Thus for the rise time τ_r and decay time τ_d of the display phenomena we have

$$\tau_r < T/n \text{ and } \tau_d > T$$

These relationships are modified somewhat if one considers the integrated effect of a number of addressing cycles but are sufficiently accurate to enable us to discuss the suitability of various effects for display purposes.

2.3   Circuits

The circuits which are used to drive a display are an important aspect in determining the economic viability of the system. Integrated circuits are widely used and in order to exploit the economic MOS range of integrated circuits the drive requirements of a display must be kept below $\sim$25V. With most liquid crystal phenomena this requirement is readily met. There are some further benefits to be obtained if the voltage required is $< 5V$ since TTL circuits can then be used. However this benefit is not very great since liquid crystal displays do not need the current levels provided by TTL circuits, and the range of available MOS circuits is rapidly expanding. However for the lowest possible power dissipation low voltage operation is desirable and devices operating at 1-2V are required.

2.4.   Summary of Basic Requirements

From the discussion of the previous paragraphs we can identify certain basic requirements that should be met by an electro-optic phenomena that is to be used in a display. Neglecting the properties that determine the visual acceptability of the display the requirements can be summarized as follows:

i.   A sharp threshold

ii.   A fast rise time ($\tau_r < 1$ msec for n = 100, T = 0.1 sec)

iii.   A slow decay time ($\tau_d > 0.1$ sec for T = 0.1 sec)

iv.   An operating voltage $< 25V$ for use with MOS circuits or $< 5V$ for use with TTL circuits

In section 4 we will see how well these requirements are met by a number of electro-optic phenomena shown by liquid crystals. However, before considering these phenomena in detail it is necessary to summarize some of the essential properties of liquid crystals.

3.   LIQUID CRYSTALS

The electro-optic phenomena shown by liquid crystals provide a potentially important method of making passive or subtractive matrix displays. The dis-

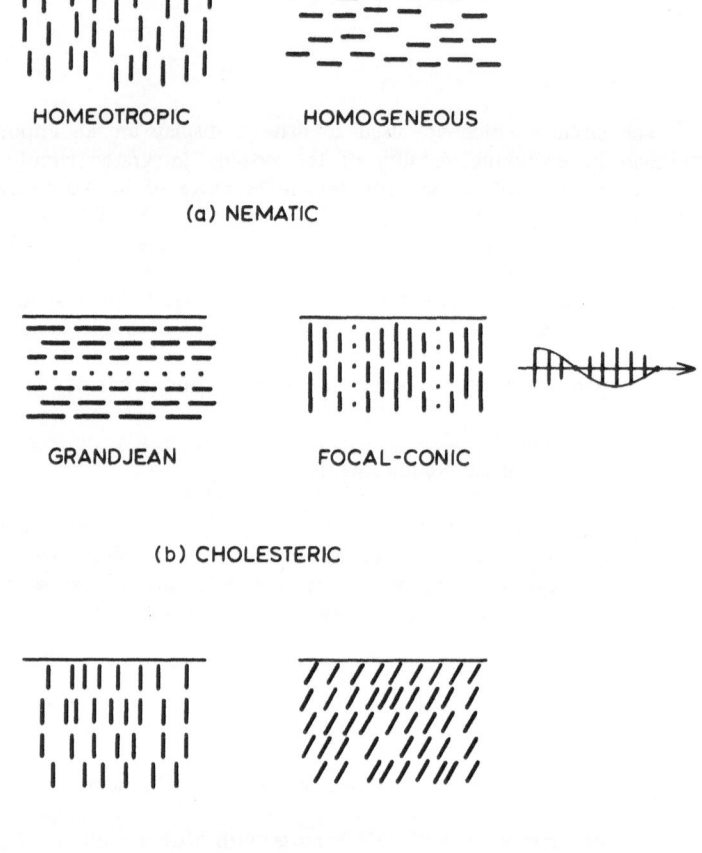

Fig. 4. Molecular arrangement of liquid crystal mesophases.

plays are made by holding a thin film of liquid crystal between transparent plates. The inside surfaces are coated with transparent electrode patterns so that voltages can be applied to selected areas of the film, as described in section 2.1. These voltages produce electro-optic effects which modulate the light incident on the device to form the display.

In this review we will not be concerned with a detailed discussion of the properties of liquid crystals. It is however necessary to briefly summarise some of the properties in so far as they are relevant to displays. We will then consider in more detail how the various electro-optic effects are used in practical devices in section 4.

3.1.   Molecular Order in Liquid Crystals

The liquid crystal state is a mesophase of matter exhibited by a wide range of organic substances. In general the molecules of these substances are highly non-spherical or rod-shaped so that various degrees of order of the molecular arrangement can be identified. This order exists only over a clearly defined temperature range, below which the material is solid and above which it is an isotropic liquid. Frequently the material will pass through more than one of these ordered states as its temperature is raised.

The molecular arrangements of the three liquid crystal mesophases are illustrated schematically in Figure 4. Of these mesophases only the nematic and cholesteric have as yet found any practical application but the smectic is included for the sake of completeness.

In displays we are usually concerned with the properties of a thin film so that it is necessary to define the molecular arrangement with respect to the plane of a film or surface. Thus we will be concerned with the following systems.

a.   Nematic
In the nematic state molecules are arranged parallel to each other, and this is the only degree of order present. In the homeotropic form the alignment is perpendicular to the surface, whereas in the homogeneous form the alignment is parallel to the surface. These alternative forms are produced by suitable treatment of the surfaces. For example the homeotropic state is obtained on cleaned surfaces or those treated with certain surface agents. The homogeneous form is obtained with surfaces which have been treated to produce a preferred direction along which the molecules lie. For example rubbed surfaces or those which have an obliquely evaporated film deposited on them can produce a homogeneous state.

b.   Cholesteric
In the cholesteric state the molecules form a screw structure which has a defined pitch and axis. When a cholesteric phase is formed by allowing an isotropic liquid to cool the focal-conic form is obtained. In this the screw axis lies in the plane of the film but the direction within this plane is undefined and varies locally so that the material is disordered and scatters light. In the Grandjean or

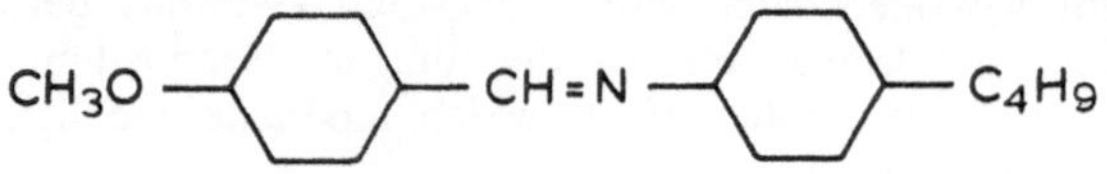

$$CH_3O - \langle \rangle - CH=N - \langle \rangle - C_4H_9$$

AZOXY COMPOUNDS   (Nematic $\Delta\epsilon \sim -0.2$)

    eg 4-methoxy-4'-butylazoxybenze

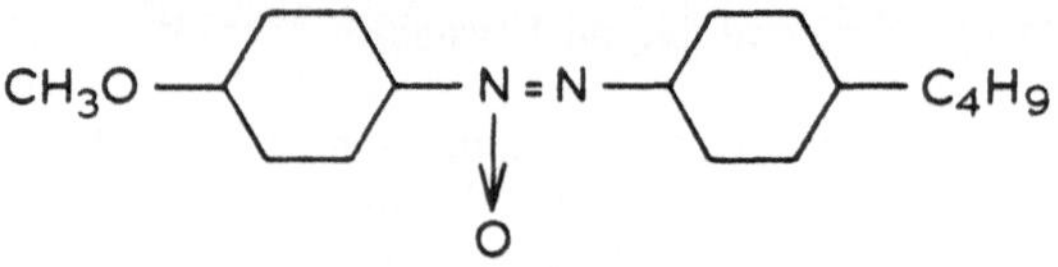

$$CH_3O - \langle \rangle - N=N - \langle \rangle - C_4H_9$$
$$\downarrow$$
$$O$$

BIPHENYL COMPOUNDS   (Nematic $\Delta\epsilon \sim +10$)

    eg 4'-n-pentyl-4-cyanobiphenyl

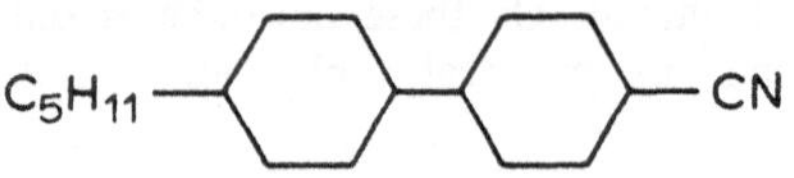

$$C_5H_{11} - \langle \rangle - \langle \rangle - CN$$

CHOLESTEROL DERIVATIVES   (Cholesteric)

    eg Cholesteryl nonanate

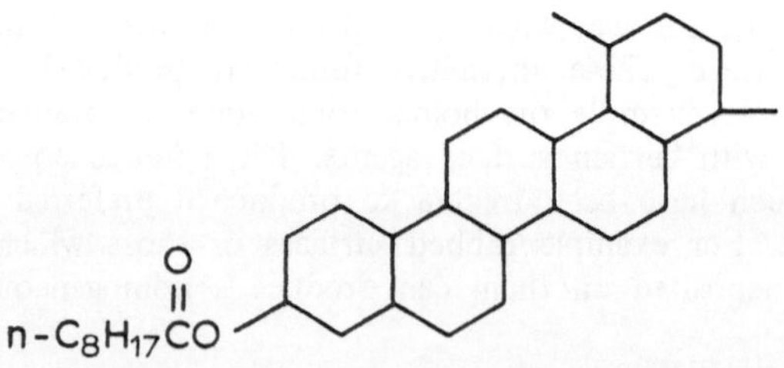

$$n-C_8H_{17}CO-$$

Fig. 5. Representative liquid crystal materials.

planar form the screw axis lies normal to the film; this may be obtained, for example, by applying a mechanical shear to the disordered focal-conic state.

c.   Smectic

The smectic phase has not as yet found use in display systems, but is included for completeness in Figure 4. Several versions of the phase occur, but essentially in each the molecules are arranged parallel to each other and in layers. The resulting smectic material is very viscous and its electro-optic properties have not been extensively studied nor applied.

3.2.   Dielectric and Optical Anisotropy

Most of the physical properties of a liquid crystal are anisotropic. For example the dielectric constant can be defined paralled ($\epsilon_\parallel$) and perpendicular ($\epsilon_\perp$) to the molecular axis. If $\Delta\epsilon = \epsilon_\parallel - \epsilon_\perp$ is positive the material has a positive dielectric anisotropy and electric field will cause the molecules to align parallel to the field. In the opposite case of negative dielectric anisotropy the molecules will align perpendicular to an electric field. Typically negative dielectric anisotropies are ~ -0.4 whereas positive values as high as 10 are common.

The refractive index of a liquid crystal is also anisotropic. It is for this reason that perturbations of a perfectly aligned state are visible. In extreme cases, as for example the disordered cholesteric state, material can be highly optically scattering, and this is the basis of a number of displays. On the other hand the optical properties of a liquid crystal can be modified by an electric field so that various interactions with polarized light are obtained. This type of phenomena is the basis of a second class of displays which will be discussed in a later section.

3.3.   Materials

In recent years considerable effort has been devoted to the search for materials which are in the liquid crystal state over a wide temperature range. The structure of some of the materials which are now being used is shown in Figure 5. Using these materials, or mixtures of related compounds, temperature ranges of typically 0 - 50°C are available.

4.   LIQUID CRYSTAL ELECTRO-OPTIC PHENOMENA IN DISPLAYS

We come now to consider in some detail the liquid crystal phenomena that have been applied to practical displays. At the present time there are five such phenomena and they can be divided into two classes.

In the first of these classes the display is formed by a transition between a clear and a scattering state of the material. The most familiar example of this class is "dynamic scattering", the other members being the "storage effect" and a cholesteric-nematic phase change effect.

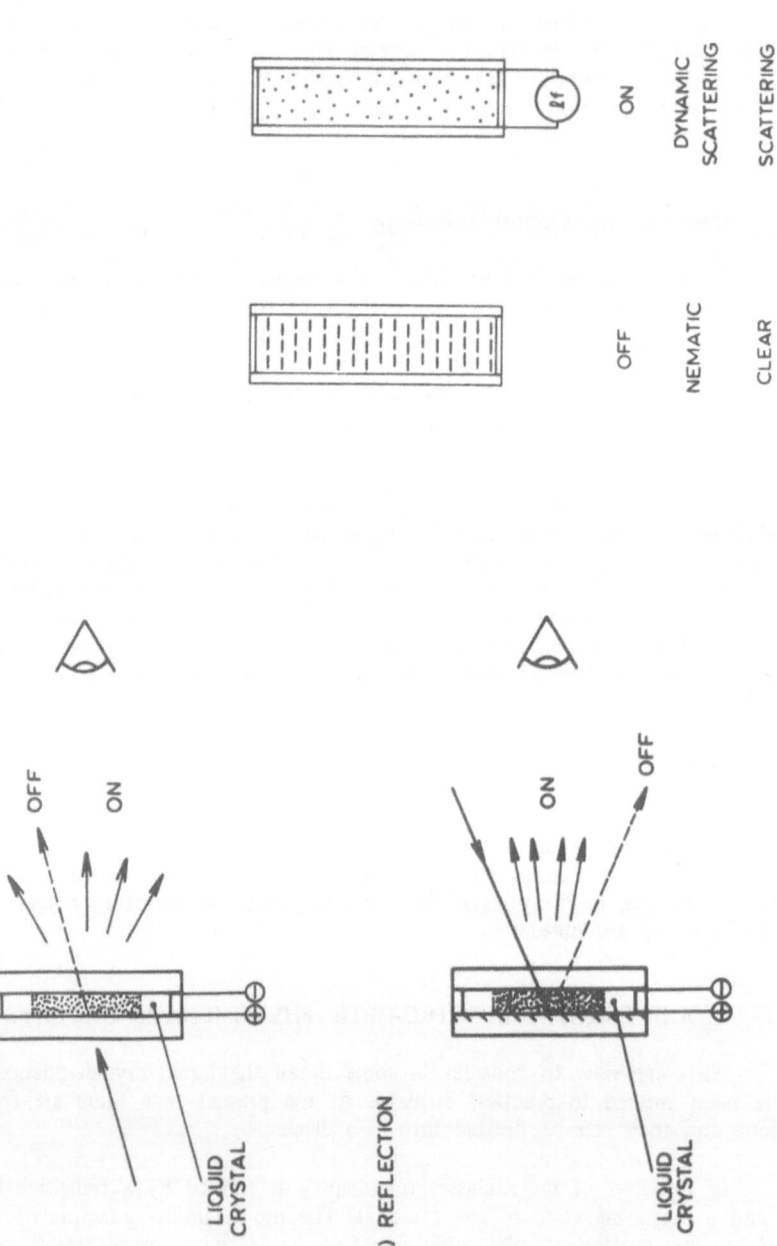

Fig. 7. The dynamic scattering ($\Delta\epsilon-$ve) effect.

Fig. 6. Schematic diagram of a light scattering display.

The way in which a scattering effect is used in a display is shown in Figure 6. As shown here both reflective and transmissive modes of operation are possible. In both of these the light source must be positioned so that it is outside the observer's field of view, and in some cases this is a severe handicap. Usually the scattering state is predominantly small angle, and thus in a forward direction, so that in the reflective display a mirror surface must be used to get intense back scattering and a high contrast display.

In the second class a change in the state of the liquid crystal interacts with polarized light. This occurs because the refractive index in a given direction can be influenced by an electric field. The change in polarization is then made visible by a second polarizer. Here the most familiar example is the "twisted-nematic" effect, the other being known as "variable birefringence".

4.1.   Dynamic Scattering (1)

Dynamic scattering is the name given to the liquid crystal effect which has been most widely applied in displays. It occurs in a nematic material ($\Delta \epsilon$ negative) when a DC or low-frequency AC current is passed through the material. The current flow induces bulk liquid movement which becomes turbulent and thus causes the material to become optically scattering as shown schematically in Figure 7.

The initial state of the material may be either the homeotropic or homogeneous form of the nematic structure. The homeotropic state is usually used since it is the more easily controlled. In this case the first effect of an electric field is to produce the homogeneous state since the molecules align perpendicular to the field. With an initially homogeneous state this first transition is not observed.

Dynamic scattering does not have a sharp threshold so that neither half- nor third-select methods of addressing give satisfactory discrimination. With typical materials a voltage of 10 - 25V DC or low-frequency RMS applied across a 25 μ film induced scattering. As the applied frequency is increased the voltage required increases, as shown in Figure 8, until a critical cut-off frequency is reached. This frequency depends on the conductivity of the material but is typically $\sim$ 1 kHz, for a conductivity $\sim 10^9 \Omega$. cm. Above the cut-off frequency bulk liquid motion and turbulence do not occur and the structure of the material is determined by dielectric effects. In this regime the material is aligned in a homogeneous state and appears essentially clear although "chevron" patterns occur at high fields.

These low- and high-frequency characteristics are used to improve the discrimination obtained with half- or third-select addressing techniques (2, 3, 4). A low-frequency signal is used to energize the display elements while a high-frequency signal is used to inhibit the scattering on partially selected elements.

The rise and decay times of dynamic scattering are of the order 10 ms and 100 ms respectively. Thus matrix addressed displays based on this phenomena are limited to systems with about 10 lines of elements.

C. H. GOOCH

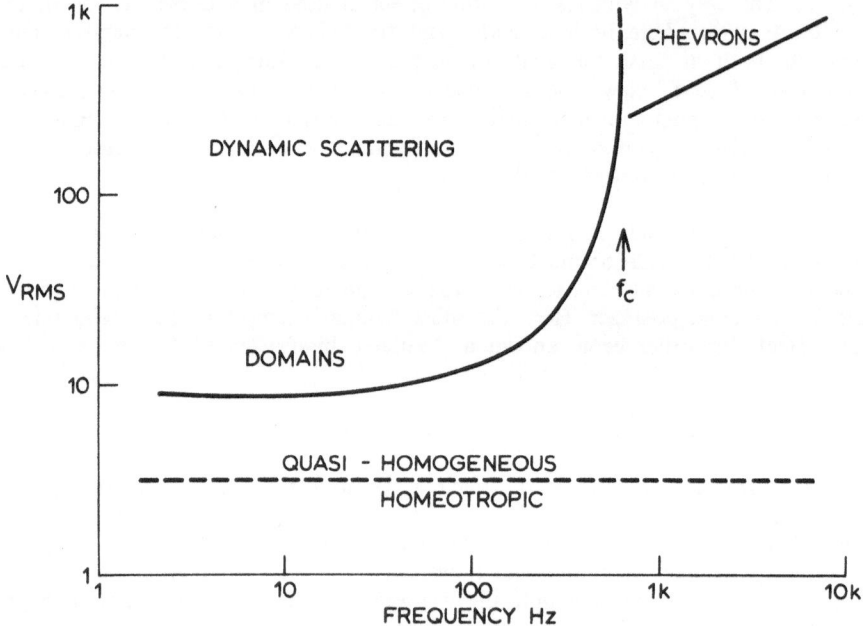

Fig. 8. Frequency characteristics of dynamic scattering.

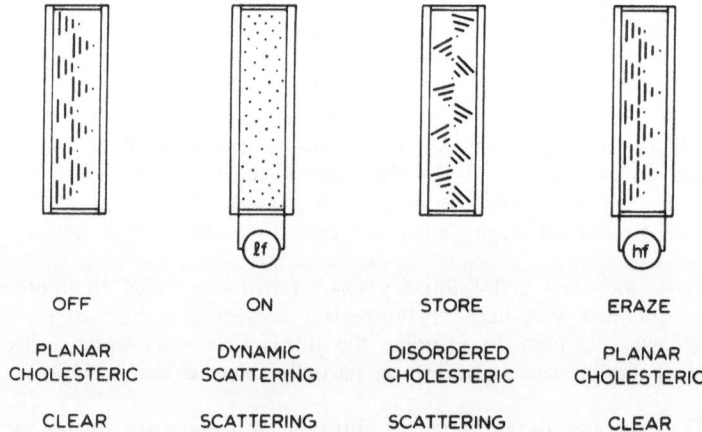

Fig. 9. The nematic-cholesteric storage effect ($\triangle\epsilon$—ve).

4.2   Storage Effect (5)

The complex phenomena known as the storage effect is illustrated in
Figure 9 and is observed in certain mixtures of cholesteric and nematic materials.
The nematic material has $\Delta\epsilon$ negative and comprises 95-98 mole % of the mixture.

In the initial state the material adopts an ordered cholesteric state and
appears clear. When a low-frequency field is applied the nematic component of the
mixture shows dynamic scattering and the material is disordered. This disorder is
retained after the field is removed so that a scattering state is stored, although the
scattering power of the stored state is not as strong as that of the dynamic scattering state. Depending upon the proportion of cholesteric material in the mixture
the stored scattering state is retained for periods extending from the natural decay
of dynamic scattering ($\sim$ 100 m sec) up to an indefinite period.

The scattering state is removed or "erased" by applying a high-frequency
field which re-establishes the ordered state.

In some respects the storage effect has properties which make it ideally
suited to a matrix display. In principle it is possible to address or write the display
in the same way as a dynamic scattering display. The information could then be
retained indefinitely until erased by a high-frequency signal. Alternatively, by a
suitable choice of mixture composition, the natural decay time can be tailored to
suit the number of display lines.

Unfortunately the advantages of the storage effect are to a large extent
negated by the voltage levels needed to erase the display. These can be as high as
100 V RMS at 1 kHz and may need 1 s or more to effect erasure. With the lower
cholesteric concentrations the voltages required are more convenient but the stored
scattering power is significantly reduced.

Thus in practice the storage effect has not been greatly used in displays
although it may be that further development will enable more useable voltage levels
to be realized. This, however, will require materials with higher negative dielectric
anisotropies than the values common at present.

4.3.   Phase-Change Effect (6)

The third scattering effect to be considered is the cholesteric to nematic
phase change effect illustrated in Figure 10. The material used here is a cholesteric
with positive dielectric anisotropy. Since the pure cholesteric materials known have
$\Delta\epsilon$ negative, suitable materials must be made by mixing a cholesteric with a positive nematic material.

In the field-free state the material adopts a disordered cholesteric structure which is scattering. When a field is applied the molecules align parallel to the
field and the material is thus in a nematic homeotropic state, which is clear. When
the field is removed the material returns to the disordered cholesteric state.

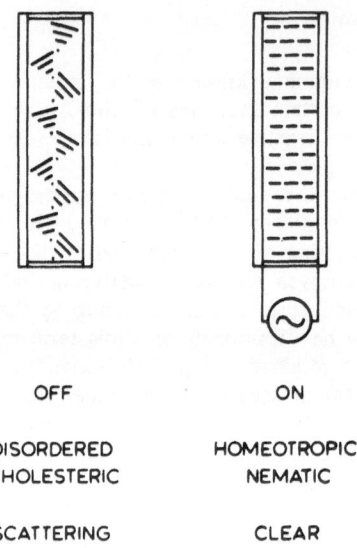

OFF   ON

DISORDERED   HOMEOTROPIC
CHOLESTERIC   NEMATIC

SCATTERING   CLEAR

Fig. 10. The cholesteric-nematic phase-change effect ($\triangle\epsilon$+ve).

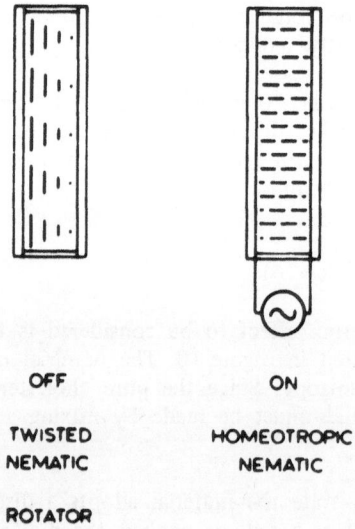

OFF   ON

TWISTED   HOMEOTROPIC
NEMATIC   NEMATIC

ROTATOR

Fig. 11. The twisted nematic effect ($\triangle\epsilon$+ve).

The transition is induced by a field, rather than by the passage of an electric current, of about 25 V RMS across a 12 μm film of material. The transition has a threshold which is sharp enough for half-select addressing techniques to give adequate discrimination. Under these conditions the observed rise time is 50 msec and the decay time is also of this order. Fortunately the decay time can be extended by applying an AC or DC bias to the display element. The nearer this bias is to the threshold value the longer will be the decay time and in practice decay times of several seconds are possible. The bias is applied continuously to all the display elements but is removed when the display is to be cleared.

Using these techniques, displays with 30 lines of elements have been demonstrated (7).

4.4   Twisted Nematic (8)

The first of the birefringent effects to be considered - the twisted nematic - is represented in Figure 11. A nematic material ($\Delta\epsilon$ positive) is contained between surfaces which have been treated to produce the homogeneous state. These surfaces are arranged so that the alignment directions at the two surfaces are mutually perpendicular so that the molecular alignment undergoes a 90° rotation between the plates.

The important property of this structure is that plane polarized light, incident with its E vector parallel or perpendicular to the initial alignment, is rotated with the molecular alignment and is thus rotated 90° by the device. When placed between parallel polarizers as in Figure 12 extinction is obtained. When a field is applied across the material a homeotropic state is obtained, because the material has $\Delta\epsilon$ positive. Optical rotation is thus destroyed and the device between parallel polarizers gives transmission. Thus the cell goes from dark to light on the application of a field, or conversely from light to dark if the device is used between crossed polarizers.

The twisted nematic effect has a relatively sharp threshold at about 5 V RMS for a 25 μ thick film and half-selection techniques give satisfactory discrimination.

Unfortunately the differences between rise time and decay times ($\sim$ 10 msec and 100 msec) are only sufficient for displays of about 10 lines of elements.

4.5.   Variable Birefringence (9, 10)

The variable birefringence effect can be obtained using nematic materials with $\Delta\epsilon$ either negative or positive. The former case is illustrated in Figure 13.

The material is initially in the homeotropic state. When a DC or AC field is applied the molecules rotate towards a direction perpendicular to the field as shown and the film becomes birefringent. The cell is placed between crossed

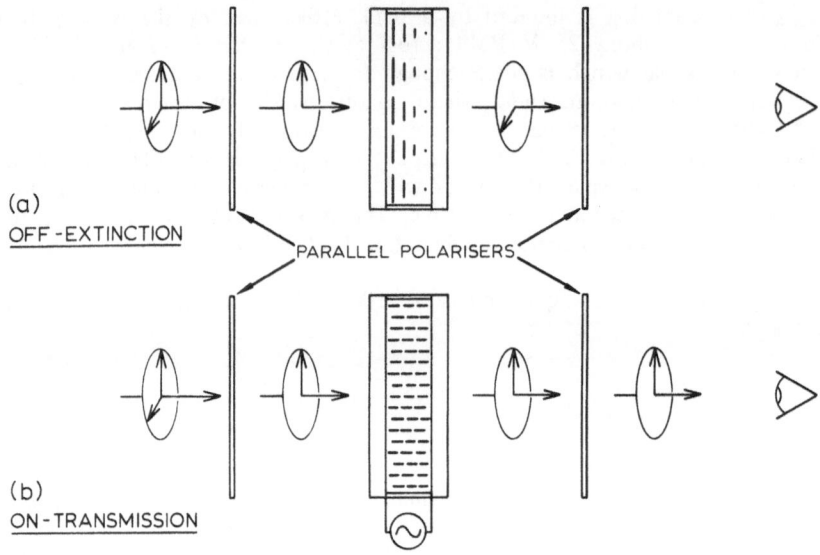

(a)
OFF-EXTINCTION

(b)
ON-TRANSMISSION

PARALLEL POLARISERS

Fig. 12. Schematic diagram of a twisted nematic display.

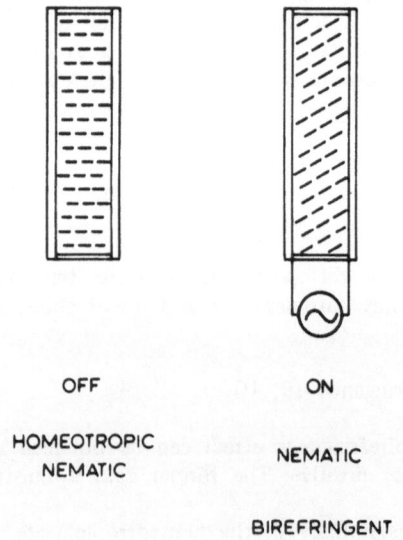

OFF

ON

HOMEOTROPIC
NEMATIC

NEMATIC

BIREFRINGENT

Fig. 13. The variable birefringence effect ($\triangle\epsilon$–ve).

Table 1. THE ESSENTIAL FEATURES OF THE DESCRIBED ELECTRO-OPTIC SYSTEMS

DISPLAY TYPE	SCATTERING			BIREFRINGENT	
DISPLAY EFFECT	DYNAMIC SCATTERING	STORAGE EFFECT	PHASE CHANGE	TWISTED NEMATIC	VARIABLE BIREFRINGENCE
MATERIAL	Nematic	Cholesteric	Cholesteric	Nematic	Nematic
$\Delta\epsilon$	−ve	−ve	+ve	+ve	−ve
OFF STATE	homeotropic or homogeneous	grandjean	focal conic	twisted	homeotropic
ON STATE	turbulent	focal conic	homeotropic	homeotropic	homogeneous
ADDRESSING METHOD	half- or third-select	half- or third-select	half-select	half-select	third-select
	Two frequency	write: ~ 100Hz eraze: ~ 1 kHz	ac bias		
VOLTAGE LEVELS	10-25V	write: 10-25V eraze: 50V	25-50V	2-5V	10-20V

polarizers and illuminated with white light. In the off state light passes unchanged through the film so that the crossed polarizers produce extinction. In the on, or birefringent, state the film is optically active and interference colours are observed. These occur because, for certain wavelengths, the induced birefringence rotates the plane of polarization by 90°, which gives transmission. To achieve a uniform effect the film must be of uniform thickness and the molecules must all tilt in the same direction. The latter condition is difficult to achieve but may be approached by treating one of the cell surfaces to give some tendency to produce a directional homogeneous alignment. This tendency must not destroy the basic homeotropic texture but merely cooperate with the field effect to define the direction of tilt. This direction should bisect the angle between the polarizers.

In order to avoid dynamic scattering the material used must be purified to have a high resistivity so that the dynamic scattering cut-off frequency is low. The display is then operated with a signal frequency that is above this cut-off.

Molecular realignment commences at a threshold field of $\sim$ 5 - 10 V across a 10 μm thick film and then increases as the voltage is increased, towards a saturation value. Half-selection techniques will give a coloured-on-black display but are usually not satisfactory as the discrimination is poor. However, variable birefringence may, with some advantage, be addressed by third-selection techniques. In this case all the partially selected elements will show the same colour, which is different from that of the fully addressed elements.

The rise and decay times encountered in variable birefringence are of the same order as those encountered in the other phenomena discussed. However, in this case the low duty cycle energisation, which occurs when a large number of display lines are scanned, changes the colour contrast of the display.

In principle this colour contrast could give an adequate discrimination over a considerable number of lines but in practice the problems of obtaining uniform film thickness and birefringence make this difficult to achieve.

5.   CONCLUSION

Some of the essential features of the electro-optic systems that have been described are summarised in the Table 1.

From the discussion of the previous sections it can be seen that none of the liquid-crystal electro-optic phenomena are ideally suited to matrix displays with large numbers of elements. At the present time dynamic scattering and twisted nematic devices have received most attention, but they are limited to $\sim$ 10 element lines. This is adequate for a number of purposes, particularly the display of numeric information.

The storage effect has considerable potential, but needs some significant improvements, in that the voltage levels required must be reduced to more acceptable levels.

With the displays based on variable birefringence the major problem is that of maintaining uniformity. This is a fundamental problem and seems likely to prevent the widespread use of this effect in displays.

Probably the most promising displays at the moment are those based on the phase change effect. The major feature here is the way in which the decay time can be controlled by means of a bias signal. This should make displays with 50 - 100 element lines possible. However, it remains to be seen if the visual appearance of these displays will be widely acceptable.

REFERENCES

(1)   Heilmeier G. H., Zanoni L. A. and Barton L. A., Proc. IEEE 56, 1162, 1968.

(2)   Stein C. R. and Kashnow R. A., Appl. Phys. Lett. 17, 51, 1971.

(3)   Wild P. J. and Nehring J., Appl. Phys. Lett., 19, 335, 1971.

(4)   Gooch C. H. and Low J. J., J. Phys. D., 5, 1218, 1972.

(5)   Heilmeier G. H. and Goldmacher J., Proc. IEEE 57, 34, 1969.

(6)   Wysocki J. J., Adams J. and Haas W., Phys. Rev. Lett., 20, 1024, 1968.

(7)   Ohtsaka T., Tsukamoto M. and Tsuchiya M., Jap. J. Appl. Phys., 12, 371, 1973.

(8)   Schadt M. and Helfrich W., Appl. Phys. Lett., 18, 127, 1971.

(9)   Kahn F. J., Appl. Phys. Lett., 20, 199, 1972.

(10)   Assouline G., Hareng M., Leiba E. and Roncillat M., Electronics Letters 8, 45, 1972.

ELECTROLUMINESCENT SEMICONDUCTOR DIODE DISPLAYS

C. H. Gooch

Ministry of Defence, Services Electronics Research Laboratory

Baldock, Hertfordshire (U. K.)

1.   INTRODUCTION

As long ago as 1907 Round, while studying silicon carbide detectors, observed semiconductor electroluminescence. However, it was not until the 1960's that useful devices became available. The laboratory devices of the 1960's are a major semiconductor product of the 1970's and are being used in an increasingly wide range of displays.

Electroluminescent or light-emitting diodes (LED's) are p-n junction devices that emit optical radiation when a forward current is passed through them. They have a number of important features which make them suitable for use as display devices. These features include:

(i)   Low power requirement. A typical device will require 10 mA at 2V and can thus be operated from a range of transistor logic circuits.

(ii)   Long life. LED's have a long life and high reliability. Typical devices operate for 10^4 - 10^5 hrs. Unlike many other luminous sources they degrade slowly rather than failing suddenly.

(iii)   High luminance. Typical LED's can be operated at luminance levels $\sim 10^4$ candela/m^2. At these levels displays can be viewed in most of the situations that are normally encountered.

(iv)   Small size. The small size of LED's make them very suitable for building into arrays to make displays.

The object of this review is to summarise the present status of LED's and provide an outline of the physics and technology that lie behind these devices.

2.   SEMICONDUCTOR ELECTROLUMINESCENCE

The band diagram of a p-n junction is shown in Figure 1. When the diode is operate in forward bias electrons cross the junction and combine with holes in the p-region. Similarly holes enter the n-region and combine with electrons, although it is normally the former process that is more important in LED's.

The energy that is released by the recombination process can, under favourable conditions, be emitted as a photon. In this case the recombination process is known as radiative recombination. Alternatively, competing processes can cause non-radiative recombination, in which the available energy appears as heat.

The maximum energy that can be emitted as a photon in a radiative recombination process is equal the to semiconductor band gap energy. If the photon is to fall in the visible region of the spectrum its wavelength must be less than 7000 Å and thus the band gap energy must exceed 1.8 eV approximately. The photon must also have a wavelength longer than that corresponding to the ultra-violet end of the spectrum. In practice this limitation is not encountered as, at present no diodes emitting in the U.V. spectral region have heen demonstrated.

We can then identify a number of requirements that must be met by a material if it is to be used in an electroluminescent device. These requirements can be summarised as follows:

(i)   The p-n junction. It must be possible to make p-n junctions in the material.

(ii)   Recombination process. The dominant recombination process must be radiative.

(iii)   Band gap. The semiconductor band gap must exceed 1.8 eV in order that visible radiations is emitted.

2.1   The p-n Junction

The techniques used to produce p-n junctions are well known in all branches of semiconductor device technology and thus warrant little discussion here.

The diffusion technique and the associated planar technology are the most widely exploited and allow complex monolithic devices consisting of arrays of individual diodes to be made economically.

In some circumstances, however, efficient devices can only be made if epitaxial junction formation techniques are adopted. Unfortunately this leads to a rather expensive device. Furthermore, since complex monolithic structures cannot be made by epitaxial techniques, arrays must be made by hybrid techniques in which individual diodes are bonded to a ceramic substrate.

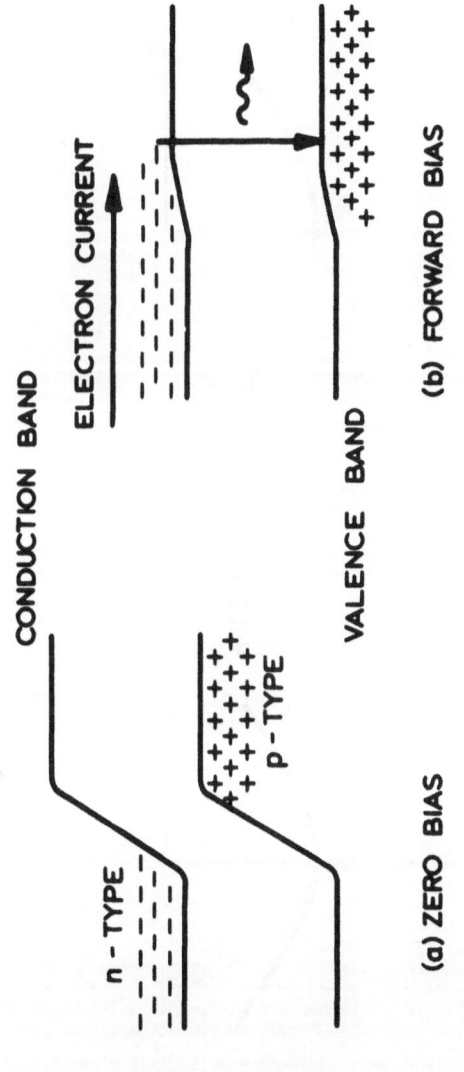

Fig. 1. Band structure of a p-n junction.

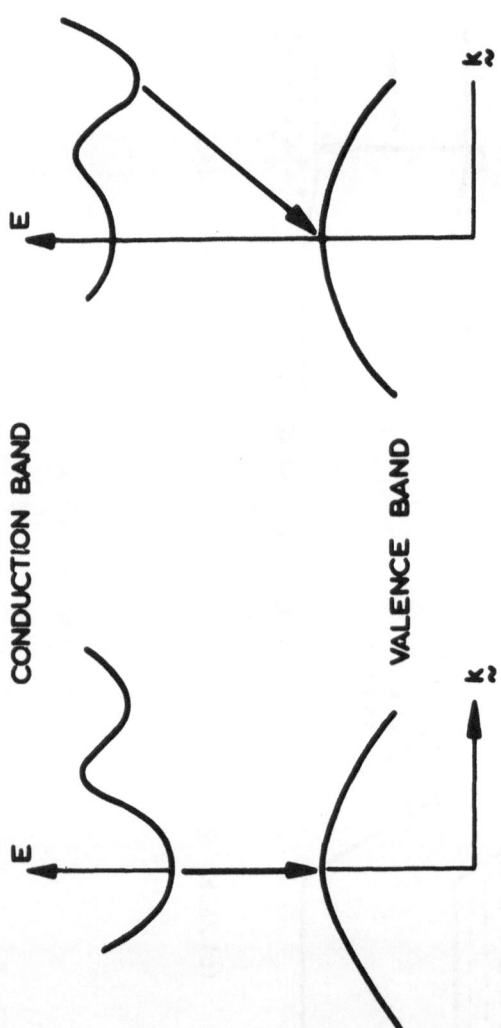

Fig. 2. Direct and indirect band gaps.

2.2   Recombination Processes

In any discussion of recombination processes in semiconductors it is necessary to distinguish between the classes of materials with 'direct' and 'indirect' band
structures respectively.

In a semiconductor the electrons and holes have a momentum k which is
a function of the electron or hole energy E (Figure 2). In a direct band-gap material
the electrons and holes at the edges of their respective bands have the same value
of momentum, whilst in an indirect band-gap material they have different momenta.
When an electron and a hole recombine they will usually do so from energy states
which are very near the edges of the bands, thus releasing an energy which is equal
to the band gap energy of the material.

In a direct band-gap material the recombination process will conserve momentum and can thus be expected to have a high probability. On the other hand,
a similar process in an indirect band-gap material would only conserve momentum if
a further interaction were involved. Thus it can be anticipated that the electron-hole
radiative recombination process will have a low probability in an indirect band-gap
material. In this case other, nonradiative, recombination processes involving impurities
or defects will be significant. These will compete with the low-probability radiative
process and may often be the dominant recombination process.

This fundamental difference between direct and indirect band-gap materials
is a significant consideration in the choice of materials for electroluminescent devices,
and leads one to consider first the direct band-gap materials. However, by paying
careful consideration to the purity of indirect materials, the probability of non-
radiative processes can be made low so that even in these materials radiative processes become dominant.

2.3   Band Gap

The sensitivity of the human eye as a function of wavelength and equivalent photon energy is shown in Figure 3. This curve has a maximum at 5500 Å
(2.2 eV) and at this wavelength radiation has a luminous equivalent of 680 lumen
per watt. For practical purposes the visible spectrum extends from about 4000 Å
to 7000 Å (3.1 eV to 1.8 eV) - at these limits the luminous equivalent has fallen
to ~ 1 lumen/watt.

We would like to have a range of electroluminescent diodes emitting over this
wavelength range, but we will see that in practice this has not been achieved and the
devices that are available are restricted to the range of wavelengths greater than
5500 Å - and extending into the infrared. In order to compare diodes emitting at
different wavelengths we must take into account the spectral variation of eye sensitivity and the emission spectrum of the diode. On this basis we can ascribe to each
diode emission a luminous equivalent ℓ . If the power efficiency of the diode is η
we can regard the product $\ell \eta$ as a figure of merit for a device - the luminous efficiency of the device. This figure of merit will be used to compare devices in the
next section.

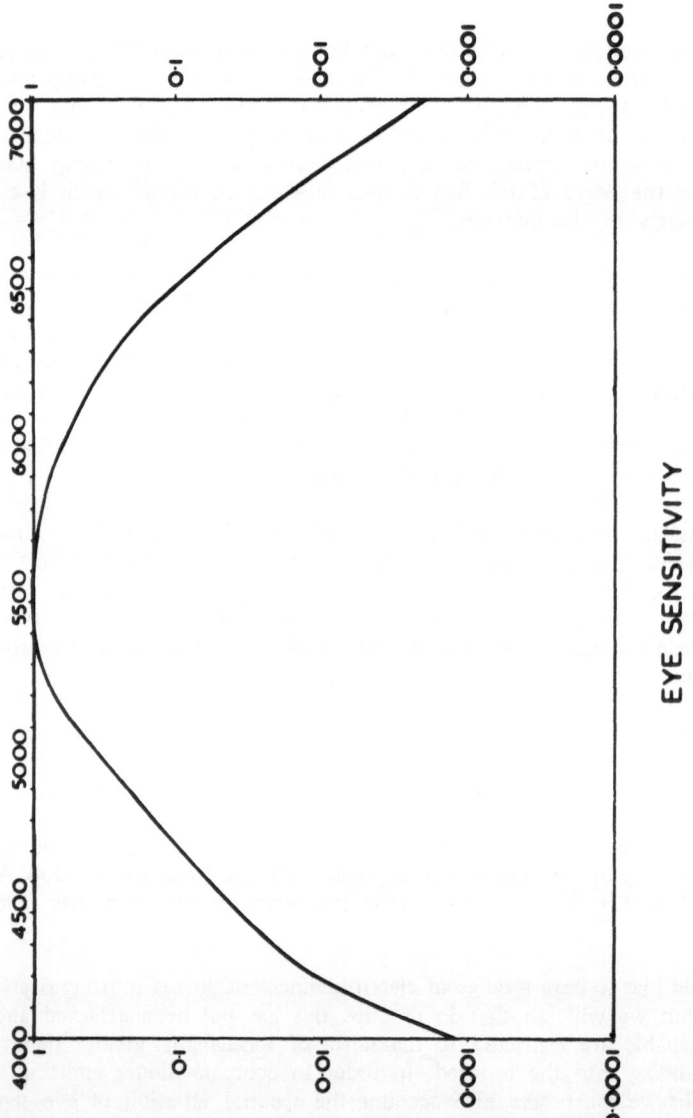

Fig. 3. The sensitivity of the human eye.

3.   ELECTROLUMINESCENT MATERIALS

The requirements of a semiconductor material wihch is to be the basis of an electroluminescent device have been outlined in the previous section. The requirements can be met in several of the III-V semiconductor compounds and at the present time these materials form the basis of all practical electroluminescent devices. However, there are potential advantages in some other materials, such as the II-VI compound semiconductors, and a considerable research effort has aimed in this direction.

3.1   The III-V Compound Semiconductors

The nine most important III-V compounds are formed from one of the elements aluminum, gallium or indium, from group III of the periodic table, and one of the group V elements phosphorus, arsenic or antimony. The compounds are all semiconductors, and various members of the family have been studied and exploited for a number of devices, besides the electroluminescent devices which are being considered here.

The band gap energies of these materials are given in table 1 and it can be seen that they range from 0.2 eV for InSb to 2.4 for $A\ell P$. This range of energies would correspond to photon wavelengths between 0.5 and 7 μm and a number of the III-V materials have band gaps which make them of interest in the context of electroluminescent devices.

(i)   Gallium arsenide

From Table 1 it can be seen that the III-V material which has the largest direct band gap is gallium arsenide. This material gives rise to efficient electroluminescent diodes but, since the band gap is only 1.4 eV, the radiation emitted has a wavelength of 900 nm, in the near infrared region of the spectrum. These diodes are thus of no interest for displays, but it may be noted in passing that it is possible to "up-convert" this radiation by using a suitable anti-Stokes phosphor to give photons of higher energy. This approach may become of greater importance if the promise of the early research work is realized.

(ii)   Gallium arsenide-phosphide

In order to make a direct band gap material that has a band gap greater than 1.5 eV one can make a ternary alloy, of the type $III_x^A III_{1-x}^B$ V or $III V_x^A V_{1-x}^B$, between two of the materials in Table 1. For example, by making the materials gallium arsenide-phosphide, with composition $GaAs_x P_{1-x}$ it is possible to make a range of materials with a band gap ranging from 1.4 eV (direct) to 2.2 eV (indirect). The transition between direct and indirect structures occurs at the composition $GaAs_{0.55}P_{0.45}$ with a band gap of 2.0 eV.

Diodes made in materials having a range of compositions in this system show a large drop in efficiency as the material changes from direct to indirect, and the emission falls further into the visible spectrum. Thus it is found that the optimum

Table 1. THE BAND-GAP ENERGIES OF THE III-V COMPOUNDS

Group III Elements	Group V Element		
	Phosphorus	Arsenic	Antimony
Aluminum	AℓP 2.4 eV (indirect)	AℓAs 2.4 eV (indirect)	AℓSb 1.5 eV (indirect)
Gallium	GaP 2.2 eV (indirect)	GaAs 1.4 eV (indirect)	GaSb 0.7 eV (indirect)
Indium	InP 1.3 eV (direct)	InAs 0.4 eV (direct)	InSb 0.2 eV (direct)

material for electroluminescent diodes has a composition $GaAs_{0.6}P_{0.4}$. These diodes emit at 6500 Å with a radiant efficiency of about 0.1%.

(iii)   Gallium phosphide

The other III-V compound that has received a great deal of attention is the indirect material gallium phosphide. The band gap of GaP is 2.2 eV, so that it should be possible to make diodes emitting at a wavelength of 5500 Å or longer. The basic problem is to overcome the inherently low radiative recombination probability in the material. This has been done, firstly, by carefully purifying the material and, secondly, by doping the material with elements that cause an interaction which enhance the radiative process.

Nitrogen-doped GaP shows a relatively high radiative recombination efficiency with an emission at 2.2 eV (5650 Å, green) for low doping levels, shifting to 2.1 eV (5900 Å, yellow) for doping levels greater than 5×10^{25} m^{-3}. The efficiency of diodes made in these materials is about 0.1%.

Material doped with zinc and oxygen also shows a high radiative efficiency at 1.8 eV (6900 Å, red), giving diodes of about 2% efficiency. This transition arises because the Zn-O doping introduces an energy level about 0.4 eV below the conduction band. The recombination process involving this level occurs into two stages, one of which gives rise to the 1.8 eV photon.

(iv)   Other III-V materials

The success of work on gallium arsenide-phosphide devices has encouraged the search for other III-V ternary alloys that would extend the available wavelength range of LED's. In particular indium-gallium phosphide and gallium-aluminum arsenide show some promise but it seems unlikely that they will compete significantly with the established materials since the potential improvements are not large.

3.2   Other materials

Although the III-V materials form the basis of all the devices that are at present available they do not offer the prospect of devices emitting at wavelengths shorter than 5500 Å. For this reason a number of other materials have been, and continue to be, studied.

Silicon carbide has received a great deal of attention. The material has an indirect hand gap of 3 eV and shows a number of radiative transitions, depending on the impurity levels present, between 4500 and 5000 Å. However, the technology of the material is very difficult and most of the research in this area has been abandoned.

Gallium nitride, which has a band gap of 3.4 eV, is also af some interest and current research is yielding some interesting results, but it remains to be seen if the material will become of any practical significance.

Table 2. THE EFFICIENCY OF LED's

Material	Wavelength, Å	Luminous Efficiency of Radiation, Lumens/Watt	Device Efficiency	
			Power, %	Luminous, mL/W
Gallium arsenide-phosphide	6500	50	0.1	50
Gallium phosphide	6900	15	1	150
	5900	400	0.1	400
	5650	600	0.1	600

The II-VI compounds are also being studied at a number of laboratories. Some of the materials in this group (e.g. ZnS, band gap 3.6 eV) make very efficient CRT phosphors but difficulties are encountered when attempts are made to form p-n junctions in the materials. Here again the ultimate usefulness of these materials remains an open question.

3.3 Comparison of Materials

The performance of the available LED's is summarised in Table 2.

From this table it would appear that gallium phosphide devices have considerable advantages over gallium arsenide-phoshide devices. However this simple quantitative comparison is not the only aspect that has to be considered and one must also take into account the different technologies involved. In pratice the technology of gallium arsenide-phosphide has been considerably further advanced than that of gallium phosphide so that most of the available devices are red-emitting and based on gallium arsenide phosphide. However the higher efficiency and colour range available from gallium phosphide make this material important in a number of instances.

The other materials that have been mentioned have not yet produced viable devices and the commercial exploitation of these materials is uncertain.

4. DISPLAY DEVICES

Although LED's are finding important applications as simple indicator lamps in this review we will concentrate on the more complex applications such as numeric and alphanumeric displays. Electroluminescent devices have properties which make them well suited to the construction of displays which contain a limited number of characters but are unlikely to become used in large-area high-resolution displays - in this area the CRT remains unchallenged.

4.1 Device Technology

Economic considerations play a very important part in determining the viability of a display device. This is because any LED display is in competition, not only with other LED displays but also with other displays devices such as gas discharge panels etc.

(a) The p-n junction

There are two alternative ways in which p-n junction devices are usually made. These techniques have been developed for a wide range of semiconductor devices and both find application in the field of electroluminescent devices.

In the epitaxial technique for forming a p-n junction a layer of one conductivity type is grown on to a substrate of the opposite type. In practice a p layer is usually grown on an n substrate, but there is no fundamental reason why the substrate should not be the p material. The p-n junction wafer which results from this growth process is then sawn into dice and some form of electrical contacts applied.

In the diffusion technique an impurity is diffused into a substrate to form the p-n junction: usually a p-type impurity, such as zinc, is diffused into an n-type substrate. One of the advantages of the diffusion technique is that it allows devices to be made by a planar technology, as is used to make a wide range of transistor devices.

The basic idea of planar technology is to inhibit the diffusion process in specified areas by means of a diffusion mask that is deposited on the surface of the semiconductor. The pattern of this mask can be defined to very close tolerances so that, for example, an array of diodes can be produced on a single substrate. The substrate can then be cut to give a number of individual diodes, or a substrate containing a number of diodes can be used as a single device, known as a monolithic device.

Epitaxial junction techiques have been mostly used to make red-emitting gallium phosphide devices whereas diffusion techniques are used with gallium arsenide-phosphide, and green-emitting gallium phosphide devices. Diffusion techniques and the associated planar technology usually lead to the more economic devices and are thus more widely used in display devices. Thus the greater luminous efficiency that has been demonstrated in red-emitting gallium phosphide diodes has not been as widely exploited in display devices as the other available systems.

(b)   Hybrid and monolithic structures

Most display devices consist of an array of LED's. There are two techniques which are used to make the required structures.

In the hybrid technique the required diode elements are bonded to a ceramic substrate. Electrical connections between these elements are made by interconnection patterns printed onto the ceramic substrate and logic circuits to drive the display can be bonded to the same substrate.

In the monolithic technique the array of diode elements is formed in a single piece of semiconductor using the planar diffusion technology. Afther the array has been formed electrical connections to the elements can be made by depositing a connection pattern over the surface of the device.

The hybrid technology is used to form the larger display devices since the individual diodes can be widely spaced with an economy in the use of material. Small displays are most easily made by monolithic techniques and character displays 2-3 mm high can be readily made.

4.2   Numeric Displays

As the present times the most widely used LED displays are numeric displays consisting of an array of 7 bars arranged in the form of a figure 8. With this format an adequate representation of the numerals 0-9 and some letters can be made. Displays of this type are used in some of the pocket calculators that are available.

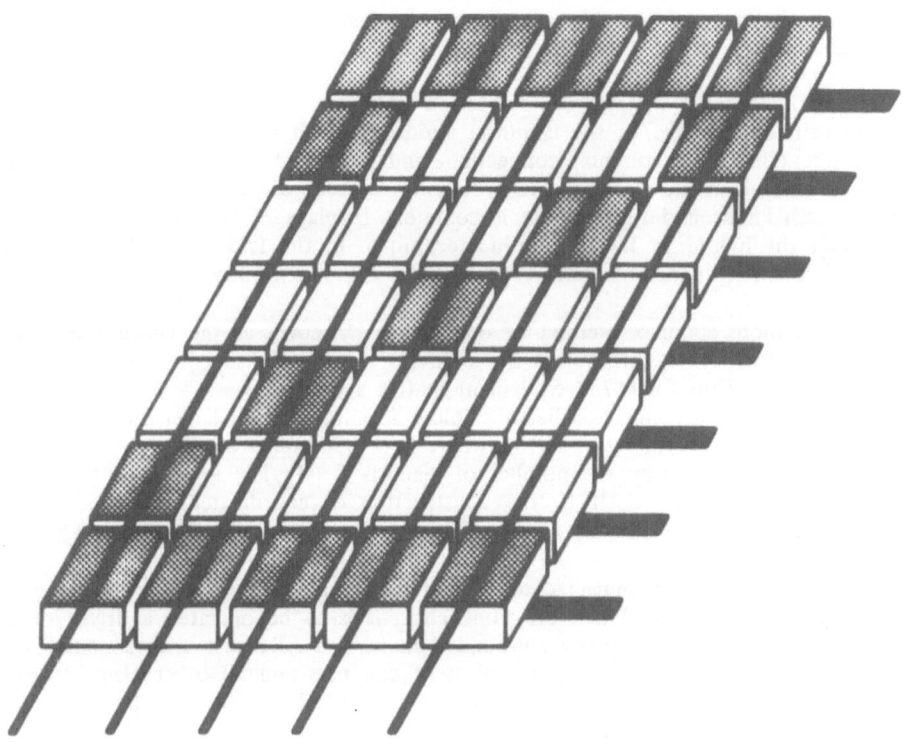

Fig. 4. The coordinate connected matrix.

4.3   Alphanumeric Displays

The 7-bar format can display the numerals adequately but if one requires the capability of displaying the full alphabet and a range of symbols it is necessary to use a more complex format. The most widely used is an array of 35 elements arranged in a matrix of 7 rows and 5 columns.

There are two basic ways in which the electrical connections can be made to a matrix display, and these have an important bearing on the way in which the devices are used.

The simplest method of connection is known as the common anode (or common cathode) display. In the common anode array all the anodes of the display elements are taken to a common connection, and individual leads are taken to each cathode of the elements. Similarly, in a common cathode display separate leads are taken to each anode and the cathode is common. Displays of this sort can be made using either the hybrid or the monolithic technique. In the latter case the substrate forms the common connection and is usually the cathode or n-type connection.

A more complex method of connection is known as the coordinate-connected display (Figure 4). In this system the anodes are connected in columns and the cathodes in rows. Thus for a 7 X 5 element matrix a total of $7 + 5 = 12$ leads are required. This method of connection becomes more advantageous for larger arrays, as the number of leads required increases as $N^{1/2}$, where N is the number of elements, whereas in the common anode array the number of leads is $N + 1$. The monolithic technology does not easily lend itself to the construction of coordinated-connected arrays since a monolithic device has, of necessity, a common connection to all the elements.

The coordinate-connected array imposes some severe constraints on the design of a display system. When only one element is to be operated a drive voltage is applied between the appropriate row and column leads. Because the elements have a diode characteristic only the required element conducts and all other elements remain off. However, if a pattern of elements is required the array must be scanned either row by row or column by column. For example, in a row-scanned display a drive is applied to the rows a, b. . . g in sequence and, at the appropriate times, a drive is applied in parallel to a selection of columns to build up the correct display. The scan must be sufficiently fast to avoid flicker.

5.   DISPLAY ELECTRONICS

A complete display system consists not only of the light-emitting display device but also the electronic circuits that drive the display. The design of these circuits plays a very important part in determining the viability of a display system and it is largely because LED's can be readily driven from semiconductor logic circuits that they can be used in displays.

We will first consider the electronics required to display a single character and then show how this can be extended to multi-character displays.

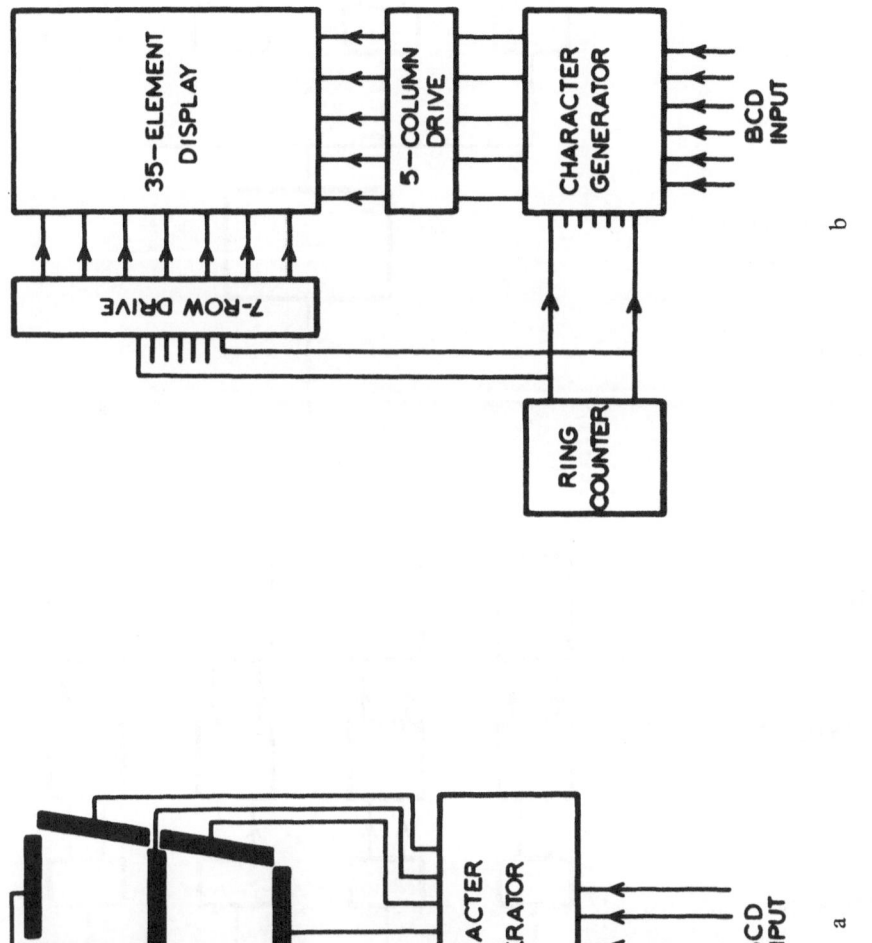

Fig. 5. The use of character generator circuits: (a) parallel output; (b) series-parallel output.

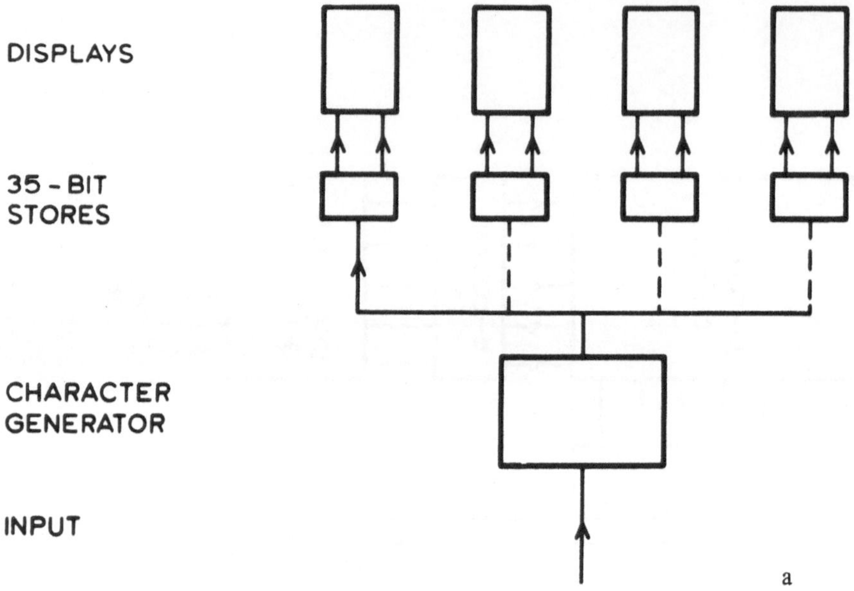

C. H. GOOCH

DISPLAYS

35 – BIT
STORES

CHARACTER
GENERATOR

INPUT

a

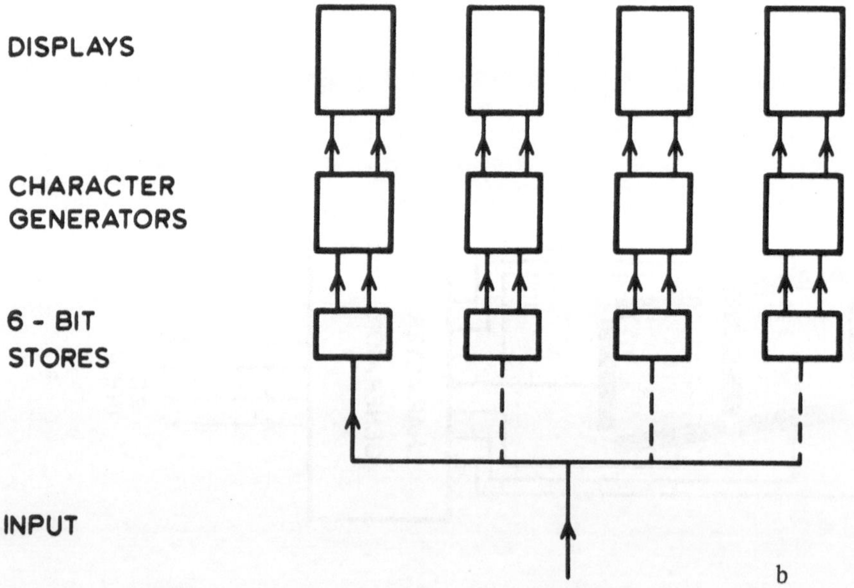

DISPLAYS

CHARACTER
GENERATORS

6 – BIT
STORES

INPUT

b

Fig. 6. The arrangement of multi-character displays: (a) shared character generator; (b) one
character generator per display.

A display will usually have a binary coded input which specifies the character which is to be displayed. In the system the following stages can be identified:

(i)   decoding circuits which identify the character;

(ii)   character generating circuits which specify which display elements are needed to form the character;

(iii)   drive stages which drive the required diode elements.

5.1   The Character Generator

It is usual to combine the decoding and character generating circuits into one "read-only-memory" (ROM) package.

For example, a typical device designed to drive a 7-bar numeral would have a 4-bit input to specify the numerals 0-9 and a 7 line output to drive the display elements (Figure 5a). This technique can be extended to cover 35 element displays, which would involve a 6-bit input and a 35 line output. In order to simplify the device it is usual to design it as shown in (Figure 5b). Here we have only 5 output lines, corresponding to the columns of a 7 x 5 matrix but two sets of input specifies the character as before: the second imput specifies the row of the row display. The operation of this type of circuit is also indicated in Figure 5b. A ring counter scans the display row by row at a rate fast enough to avoid flicker and also drives the character generator through a 7-row cycle. The output lines are connected to the display columns so that the character corresponding to the bcd input is built up.

5.2   Drive Stages

Since an LED has a low impedance diode characteristic the supply to each element must be current stabilised. For the 7-bar and similar common anode displays this presents no problems. For a matrix display the situation is more complex. Each row lead may have to drive up to 5 elements in parallel, but for only a 1 in 7 duty cycles. Since the number of elements is undetermined this drive must have a low impedance. The column drive operates only one element at a time and must be current stabilised. This drive may be on at each stage of the scan cycle. These features must be considered when considering the current and power handling capability of the drive circuits.

5.3   Multi-character Displays

Almost all displays contain a number of characters, so that it is necessary to consider how the circuits used to address a single character can be adapted to meet this requirement. In principle, the alternative approaches involve either sharing one character generator between a number of characters, or providing a character generator for each character. The choice between these alternatives is often determined by economic condiderations and the relative cost of the various components of the display system. For example, if character generator circuits are expensive it will be advantageous to use one character generator to drive a number of characters (Figure 6a).

If the character generator is inexpensive it might be more economical to provide a character generator for each display character (Figure 6b). This approach is sometimes adopted by including the character generator circuit in the package containing the display device.

Either of these approaches requires the use of storage elements in the display. In the first case the character generator specifies 36 bits of information for each display and these must be stored by a 36-bit memory associated with each display.

In the second case the bcd input is stored by means of a 6-bit memory for each character.

CONCLUSIONS

The display devices that are now available range in complexity from simple indicators to multi-character alphanumeric devices. Most of these devices emit in the red spectral region but this is being extended to give green and yellow-emitting devices.

The range of logic circuits needed to drive LED displays is also expanding and the cost of both display and logic devices is falling so that the application of LED displays is expected to continue its expansion.

GAS DISCHARGE DATA DISPLAYS

G. F. Weston

Mullard Research Laboratories, Redhill, Surrey (England)

1.   INTRODUCTION

The study of gas discharge phenomena dates back to the middle of the nineteenth century, with lamps, rectifiers and other discharge tubes appearing early in the twentieth century. It is a little surprising, therefore, with the rapid progress in electronics and electronic physics over the last twenty years, especially in the solid state physics area, that we are today still considering gas discharge devices for future applications, and especially as the modern competitor to the cathode ray tube.

The reason gas discharge devices remain a challenge to alternative methods of display stems from two important parameters of the discharge. The first is the relatively efficient production of light and the second is the bistable characteristics resulting from the ignition and extinction voltage thresholds.

The type of discharge which has been extensively exploited for display output of electronic equipment has been the neon glow discharge. Neon has a higher light output efficiency than other gases, of the order of 0.5 lumens/Watt, which gives a clearly visible display in daylight at modest current densities. Although requiring a supply voltage greater than 100 volts, the thresholds of the neon glow discharge are such that it can be switched on or off by voltages and currents compatible with solid state circuit technology. The discharge can be positively located by the electrodes, and in general the display devices can be simple in structure and thus inexpensive. Finally, the devices are reliable, with operational lives of tens of thousands of hours. There are, of course, disadvantages, the major one being the time taken for the establishment of a discharge and for its decay. This limits its switching speed to several microseconds, or even milliseconds in some circumstances.

However, with a balance of desirable characteristics it is not so surprising after all that the application of glow discharge tubes to in-line digital read-out has

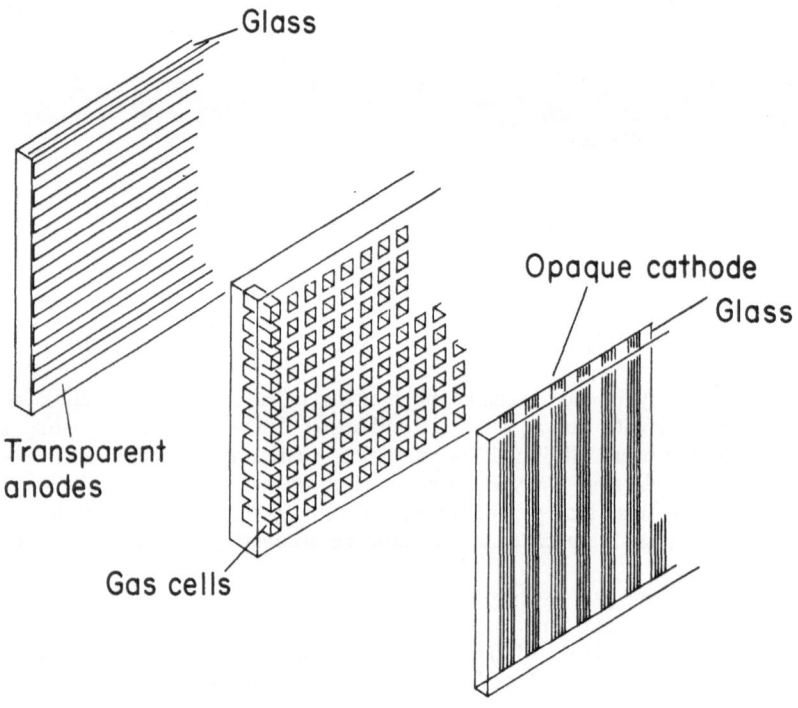

Fig. 1. Exploded view of a cross-bar gas discharge panel.

been so successful, and that with the growing need for more sophisticated displays, it has become a major area of investigation. These investigations have culminated in the development of gas discharge panels for displaying alpha-numeric data or even graphic, challenging the highly developed rival cathode ray tube in the area of high resolution display. In this paper such panels will be described and the current state of the art reviewed.

2.   GENERAL CONSIDERATIONS OF DISPLAY PANELS

The gas discharge panel consists of a two-dimensional array of discrete gas discharge which can be selectively established by a cross-bar addressing system to display the required information in a dot format. In many designs the discharges are confined in separate cells formed by an array of small apertures in an insulating plate which is placed between two electrode systems of parallel wires or strips mounted orthogonally on further insulating plates. An exploded view of such a panel is shown diagrammatically in Figure 1. The three plates are sealed together enclosing a gas mixture which is predominantly neon at a reduced pressure.

The panel can be operated under a.c. conditions in which each set of electrodes acts alternately as anodes and cathodes, or under d.c. conditions in which one set of electrodes acts permanently as cathodes and the others as anodes. In the a.c. operation the electrodes need not be in contact with the ionised gas, but may be isolated from it by an insulating layer.

To initiate the discharge in the required cell a potential greater than the ignition voltage must be applied across it. To ensure that unwanted cells are not also ignited, co-incident pulses, positive to the anode and negative to the cathode, are applied to the appropriate row and column electrodes, such that the combined amplitude is greater than the ignition voltage, but a single pulse is not, Figure 2. Since it is impossible to apply coincident pulses to two cells in different rows and columns simultaneously without coincident pulses also appearing across unwanted cells, the cells must be addressed sequentially either a cell at a time or a row or column of cells at a time. The latter, referred to as line dumping, allows a faster address time, but does not allow random access.

Two drive modes of operation for both a.c. panels can be identified:
a)   Storage operation in which each cell remains "on" after it is addressed
b)   Cyclic or sequential operation in which each cell is illuminated only during the addressing pulse period.

The brightness of the panel depends on the mean current and therefore, although the mean power requirements of both modes will be similar, the peak current requirements in the cyclic mode will be higher than for the storage mode because of the duty factor. High peak currents can cause deterioration of the cathode by ion bombardment and this imposes practical limitation in the cyclic mode on brightness and/or panel size. There is also a limitation on drive circuits, especially if the cyclic mode is line dumped. For then the row switching device has to provide the total peak current for all the selected cells in the row. Although there

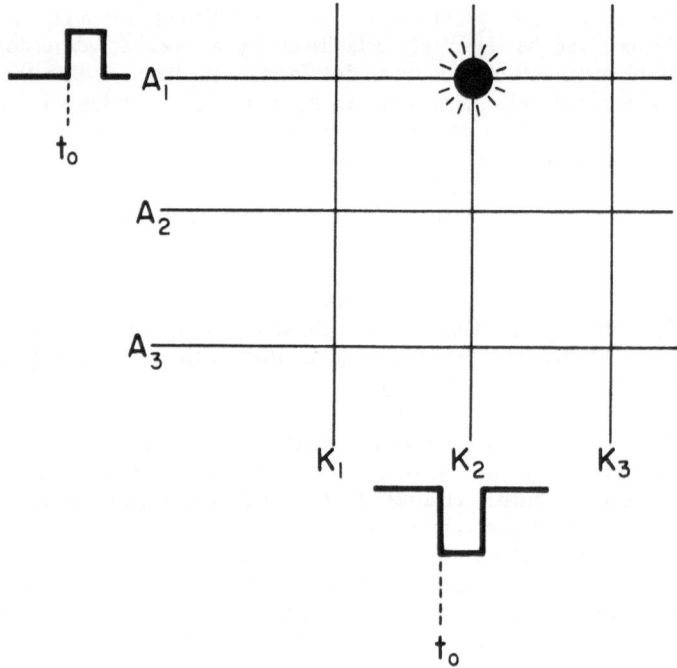

Fig. 2. Schematic diagram of cross-bar address with two coincident pulses.

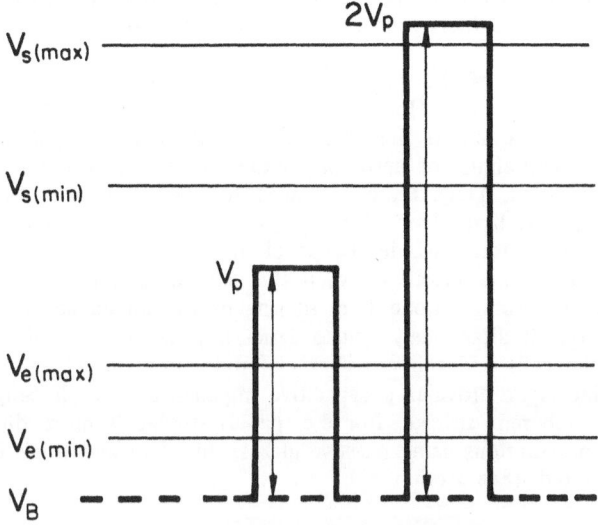

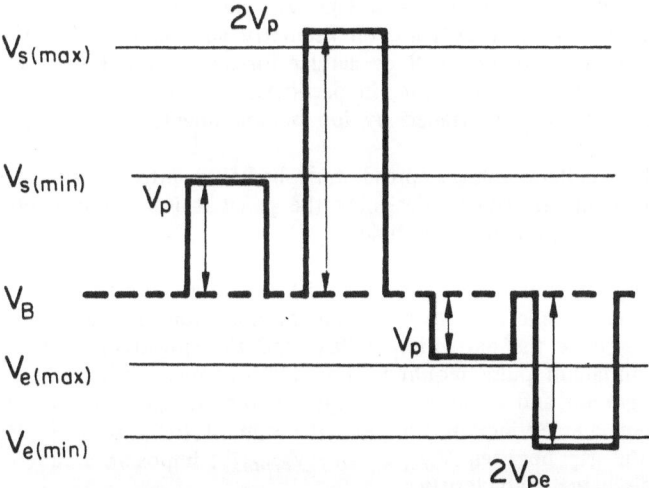

Fig. 3. Drive pulse voltage requirements of a d.c. panel. Panel functioning in the cyclic mode (top) and storage mode (bottom).

is no theoretical limit, in so much as several devices could be used in parallel to drive the row, there are strong economic reasons for limiting this current. In general these limitations restrict the cyclic operation to panels of 256 x 256 lines.

The mechanism of the a.c. and d.c. driven panels are rather different, and will be discussed separately. However, there is one other common characteristic which can be pointed out at this stage. Once a discharge is ignited the voltage drops to the lower maintaining or sustaining value which is substantially constant over a wide current range. Thus, there is a need to provide a current-limiting impedance in series with each discharge for any practical device. In the cyclic mode, the impedances can be placed outside the panel in series with one set of electrodes, each impedance is then time shared between the cells along the electrode to which it is connected. In the storage mode time sharing of the impedance is impossible since the cells are on simultaneously, and a series impedance is required for each cell, i.e. between the cells and the cross-bar electrode. For the a.c. panel the inclusion of an insulating layer provides a capacitive impedance to each cell, and the panel has therefore inherent storage. For d.c. panels storage is more difficult to achieve, although constructions using resistive glazes, thin film and thick film resistors have been demonstrated. (See section 3.3.)

3.   D.C. PANELS

The characteristic voltages of a cold-cathode glow discharge, namely the ignition voltage V_s, the maintaining voltage V_m and the extinction voltage V_e, depend on cathode and gas properties, on geometry and on pressure. Therefore, for any given panel these voltages will be similar for all cells, but not identical. Also, variations can be expected over life, in particular the cathode can become contaminated by gas impurities, or cleaned by ion bombardment.

The resultant voltage spread must be taken into account when considering the circuit requirements, and in designing the panel it is obviously advantageous to reduce the spreads as much as possible.

Figure 3 shows the pulse and bias potential requirements in relation to the spread for the cyclic and storage mode of operation. In the cyclic mode the panel is biased below the extinction voltage and the spread in ignition voltage determines the minimum pulse required. It is advantageous in this case to have a small gap between ignition and extinction voltage. In the storage mode the panel is biased between ignition and extinction and now the sum of the spreads in V_s and V_e must be less than the gap between $V_{s(min)}$ and $V_{e(max)}$, imposing a much tighter tolerance on the discharge characteristics.

When operating under pulse conditions a further factor has to be taken into account, and that is the temporal growth and decay of the discharge, i.e. the ionization and de-ionization times. They not only limit the rate at which the panel can be addressed, but they also affect the voltage requirements. In general, the shorter the pulse the higher is the amplitude required to initiate the discharge; typically to switch a cell within 10 μsec requires a voltage 25% higher than the d.c. ignition

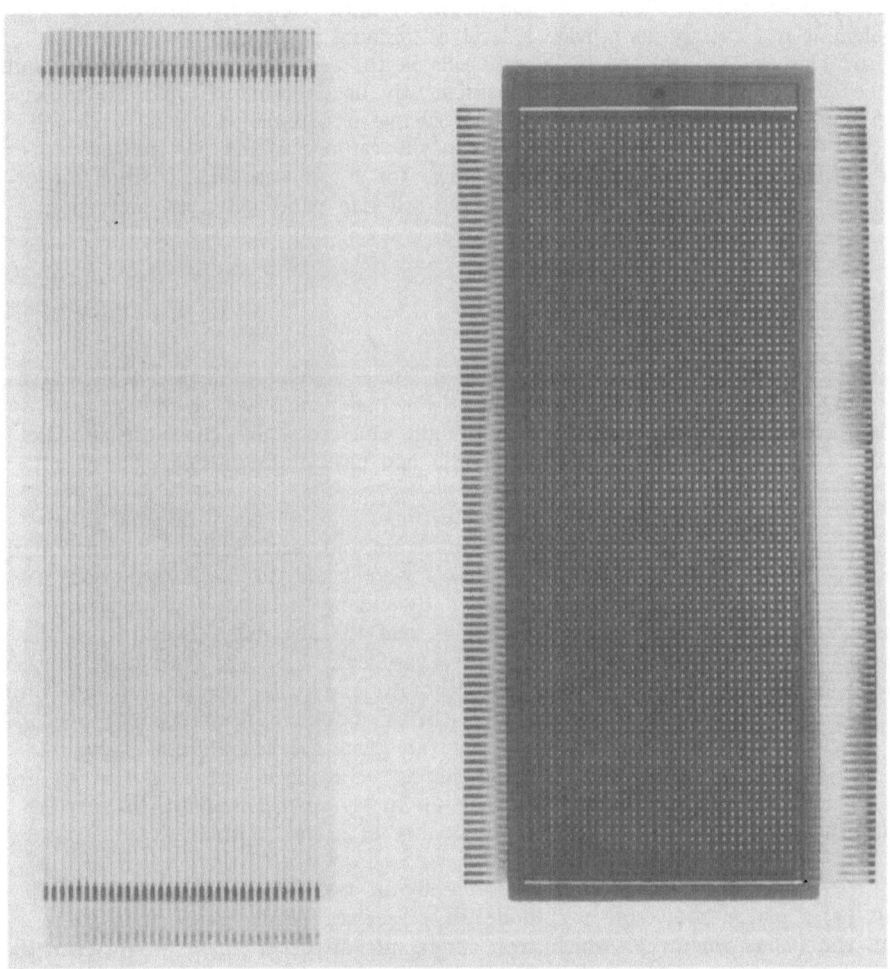

Fig. 4. Components of a d.c. panel constructed by a moulding technology.

voltage value. On the other hand, the cell can be "primed" by the presence of charged particles in the vicinity, in which case the required pulse amplitude and/ or width will be reduced. The priming may be provided by an adjacent discharge or as a result of incomplete de-ionization, resulting from a previous discharge in the same cell. Thus, not only are the voltages higher for pulse operation, but also the spread is increased due to the fact that some cells will be isolated, whilst others will be close to "on" cells and heavily primed. Generally, to overcome this problem it is necessary to provide a level of priming in all cells to reduce the spread. This can be achieved by having cells in the array, such as rows or columns at the edge of the panel, which are permanently on. A more effective method has been described by Holtz (1), whereby a discharge is maintained behind each cell which is to be ignited, and a small hole in the cathode allows the charged particles through. In this way the ignition voltage for a cell was reduced from 250V to about 160V. This method is used in the self-scan panel described in section 3.2.

3.1   Cyclic Panels

From the above we can see there are a number of factors which have to be taken into account when designing the actual panel, such as voltage and current characteristics, lumen-efficiency, priming and cross-talk. There are also problems of cathodic sputtering, operational life and cost of fabrication.

However, within these constraints there are still many possible panel constructions and technologies which can be employed, each having its merits. Early panels developed in the U.S.A. in the 1960's were based on the construction of Figure 1, with the aperture plate fabricated by etching holes in a photo-sensitive glass which is later fired to an opaque state, and the electrodes deposited on the glass plates (see for example the panels of Lear-Siegler Inc.) (2). Two basic designs have been pursued in the Philips-Mullard laboratories (3). In one the cells are formed as an integral part of the cathode array by a moulding technique (4) and in the other the cells are formed with an aperture plate fabricated from aluminum which is subsequently anodized (5). The moulding technique is particularly suitable for small panels, but probably limited to panels up to say six-inch square. The aperture plate panel on the other hand has the potentiality of extension to much larger panels.

The former panel consists essentially of two moulded glass-metal components (1) a planar body with a cathode array recessed in the glass and a pump stem, and (2) a window to which strip anodes are attached, Figure 4. An isometric view of part of a typical panel is shown in Figure 5, where the detailed structure of the cathode plate can be appreciated. It will be noted that each row of cells is isolated by ridges on which the window is supported. This prevents cross-talk between cathode rows whilst allowing priming along the anode direction. The aperture plate panel employs a 0.2 mm aluminium sheet which is etched to give the necessary structure and anodized for electrical insulation with an oxide layer of about 50 μm. The resulting anodized plate has a thermal expansion coefficient which matches that of the glass.

Window

Isolating
ridge

Anode
strip

Glow
discharge
cell

Recessed
sputter
trap

Cathode
strip

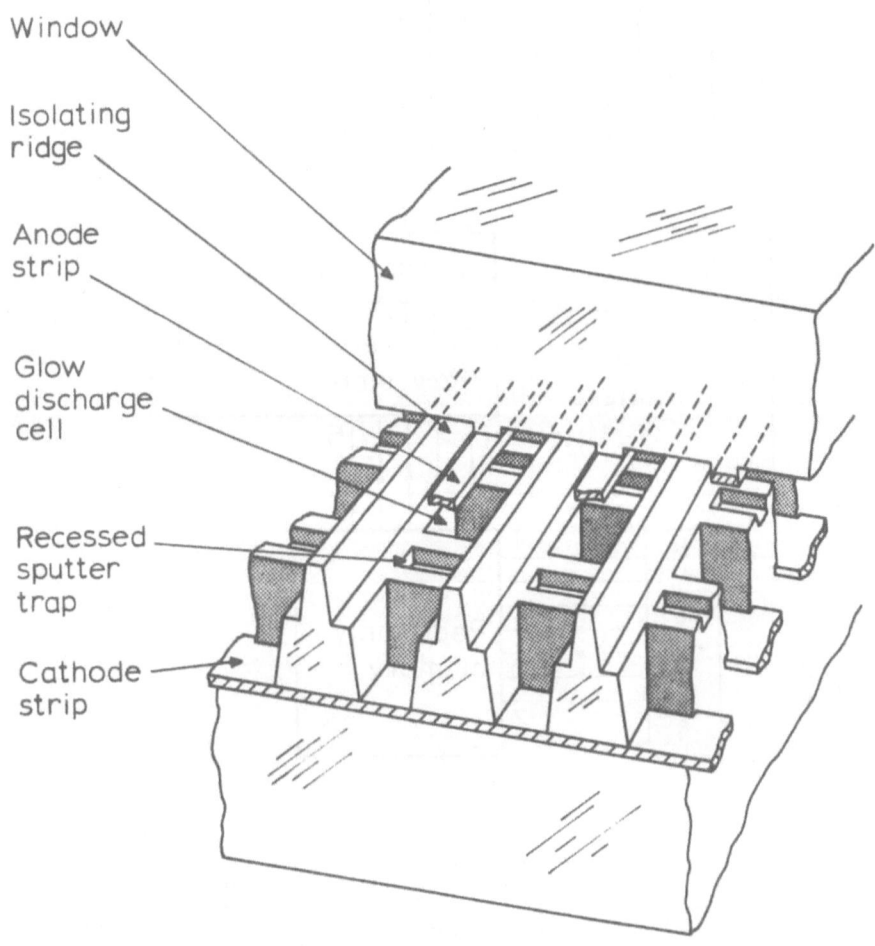

Fig. 5. Isometric view of part of the moulded panel.

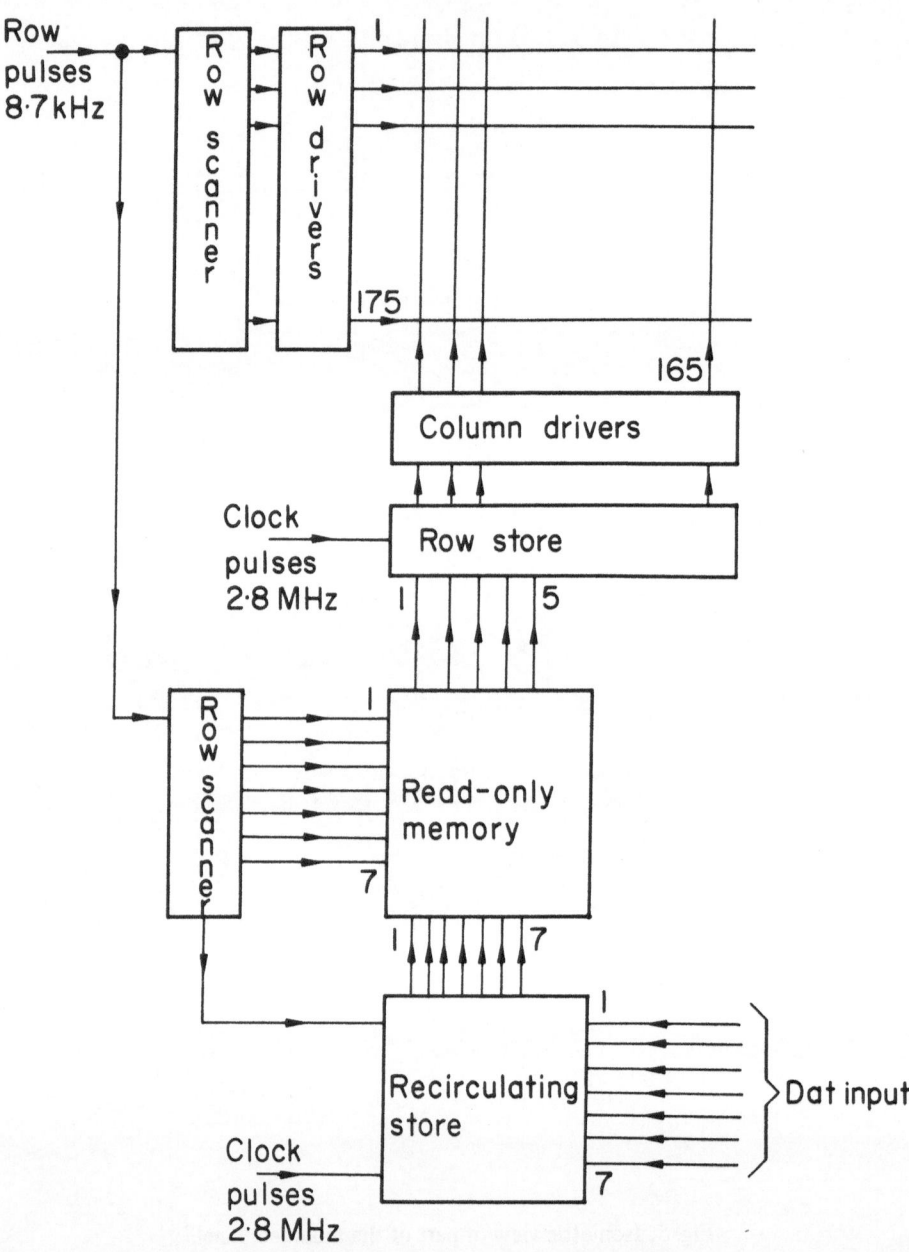

Fig. 6. Block diagram of the addressing circuit for a tabular display using a d.c. panel in the cyclic mode.

Fig. 7. A 56 character panel, cyclic addressed.

Fig. 8. Rear view of the 56 character panel showing the drive circuit construction.

Cyclic panels are of particular interest for tabular data display, where up to a 1000 characters are required. Such a system has been described by Jackson and Johnson (6) and is shown schematically in Figure 6.

Each anode row is addressed in sequence and, during the row address period, information relevant to the lighted points in that row is held in the row store and used to give simultaneous address of appropriate columns. To avoid flicker, each row is re-addressed at intervals of at least 20 msec. Input data from, for example, a keyboard, consists of a series of six-bit words in standard (ASC11) codes describing the character to be displayed. This is fed into the buffer memory, comprising of a re-circulating MOS shift register, which holds the information for the complete panel, and is used to refresh the display. Data from this store is fed to the character generator, a MOS read-only-memory, each six-bit word defining a character in the memory. The character generator is scanned in synchronism with the row scanner. At the start of each field period the first line of characters is read from the re-circulating store and the first row of each character fed into the row store. When the row store is filled the anode (row) is switched on and the first row displayed. In the next period the second-row is displayed and so on until the whole panel has been scanned.

Figure 7 shows a 56-character panel using the moulding technique mounted on a standard keyboard. A feature of this system was the attention paid to circuit packaging and interconnections. The scanning circuits, row and column drivers and the row store are all mounted on printed circuit boards at the rear of the panel with flexible film connectors, Figure 8. Integrated circuit drivers have been constructed to drive 7 rows and 5 columns, to reduce the circuit package.

Figure 9 shows an aperture plate panel displaying 112 characters operating in a similar system.

3.2.   Self-Scan Display

The cross-bar panels described so far require a switch on each electrode (i.e. n+m switches for mn elements) normally requiring 60 - 80 V transistors. The drivers alone, therefore, constitute a significant cost in the display system. A design of cross-bar display is commercially available from Burroughs under the name of "Self-scan" which reduces the number of drivers required in one direction to three, thus effecting considerable circuit economy (7). The panel on the other hand is more complex, having three sets of electrodes. An exploded view of the panel is shown in Figure 10. The panel can be thought of as having two sections: (1) the "glow-scan" section, which consists of the scan anodes and the rear side of the cathode conductors, and (2) the display section, consisting of the display anodes, and the front side of the cathode conductors with an insulating aperture plate in between, similar to the conventional d.c. panel. The two sections are linked via the small glow-priming apertures in the cathodes.

In operation, a glow is transferred down the length of the panel at the rear of each cathode, at a field rate of approximately 60 Hz. It is hardly visible

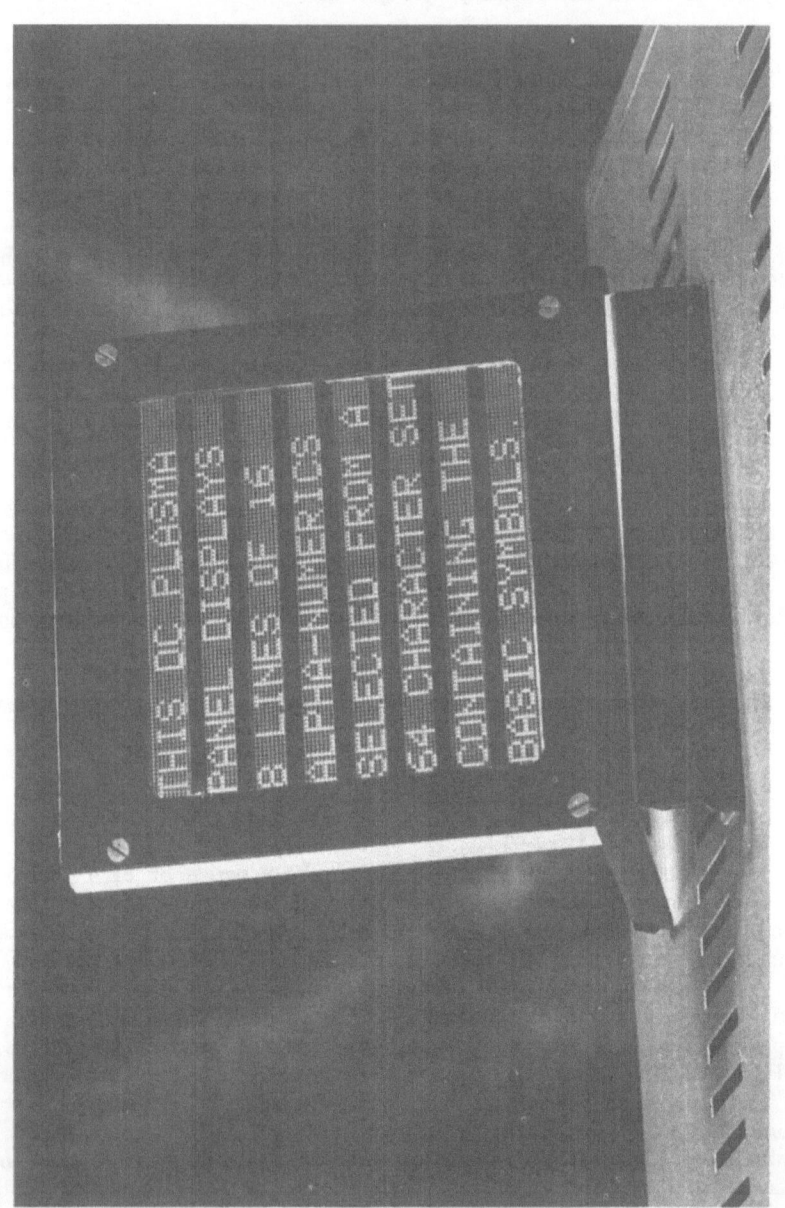

Fig. 9. 128 character panel using an anodised aluminum aperture plate.

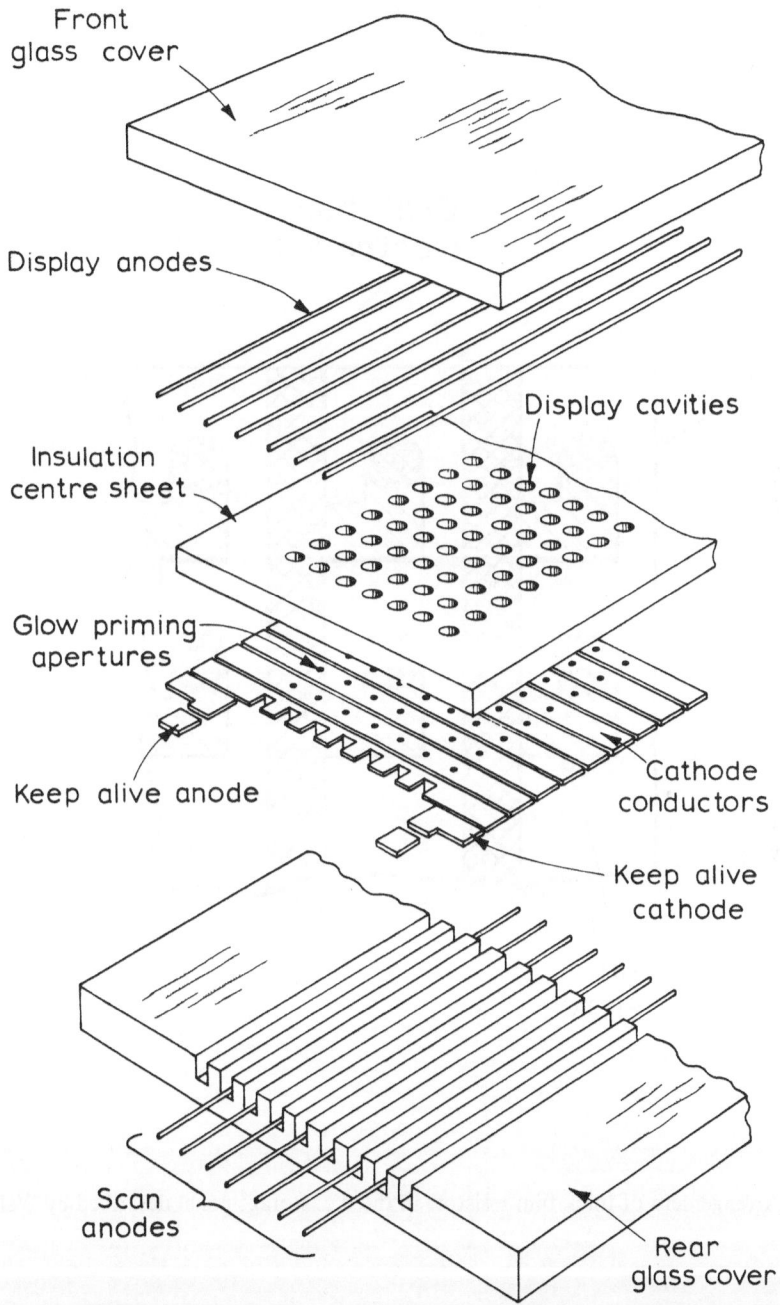

Fig. 10. Exploded view of a section of the 'Self-Scan' panel[7].

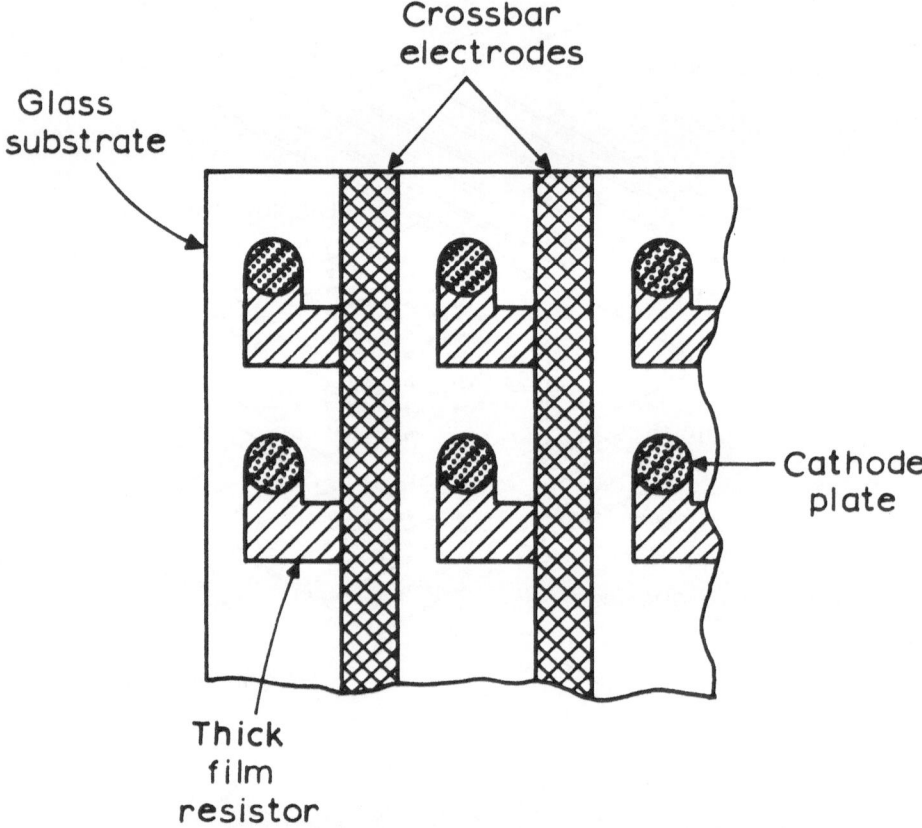

Fig. 11. Arrangement of thick film resistors in the d.c. storage panel described by Walters[9].

from the front since the cathode apertures are very small. As the glow occurs at the rear of each cathode strip, it primes the cells of the display section in front of that cathode strip, reducing their breakdown potential as described by Holtz (1). If, therefore, a positive pulse is applied to one of the anodes at the same time, of such an amplitude that it will ignite a primed cell but not an unprimed cell, then a visible glow will occur in the cell associated with the designated cross-point. Thus, by parallel addressing the anodes in synchronism with the glow transfer, the desired information can be displayed, at a field rate of 60 Hz and with a duty cycle of 1:n, where n is the total number of cathode bars. The transfer of the glow along the back of the cathodes is effected by the application of three clock pulses and can be compared with the methods used in glow discharge counting and stepping tubes. Every third cathode is connected in parallel, and 100 V drive pulses are applied in sequence to the three bus bars so formed. The scan anodes are connected to a 250 V supply via resistors R_a. The amplitude of the pulses are such they will only ignite the heavily primed cathode adjacent to the lit cathode. The extra current taken through R_a on igniting the cathode lowers the anode voltage and extinguishes the glow on the previous cathode. The fact that adjacent cathodes are heavily primed and have lower ignition voltages to the rest of the cathodes ensures that the glow progresses a step at a time in one direction with the application of sequential pulses. Panels which can display up to 256 characters are commercially available. For the larger panels a six clock pulse system is employed.

3.3   Storage Panels

As pointed out, the duty ratio restricts the cyclic panel to about 1000 characters. To display a larger amount of alphanumeric information or graphics, storage in the panel is essential. Several approaches towards a d.c. storage panel have been made and one method has reached the stage of development. Early attempts to integrate a resistor into each cell depended on coating the anodes or cathodes with a resistive coating such as conducting glass, as for example in the panel described by de Boer (8). Indications, however, suggest that the methods ran into problems of uniformity and tolerances and were later dropped; a resistor of 2 - 5 MΩ is required in each cell.

An alternative method recently described by Walters (9) seems more promising. It uses a standard technique which is becoming popular for circuit construction, namely thick film printing. The anode and aperture plate are similar to that used in cyclic panels, Figure 1, but the cathode plate incorporates an L-shaped resistor in each cell. This is illustrated in Figure 11. One end of each resistor is connected to a cathode cross-bar conductor which is offset from the cell cavities. At the other end a nickel cathode plate is deposited. A covering glaze is printed over the whole plate except for the cathode areas which are exposed. Walters quotes results for a cell pitch 1.25 mm centre to centre, but claims that double this resolution can be achieved. The brightness of the panel is high, 3,300 candelas/m^2 with a power dissipation of 60 mW per character.

Another method of forming resistors has been employed by Smith (10) of our laboratories using evaporated thin film techniques. As with the thick film

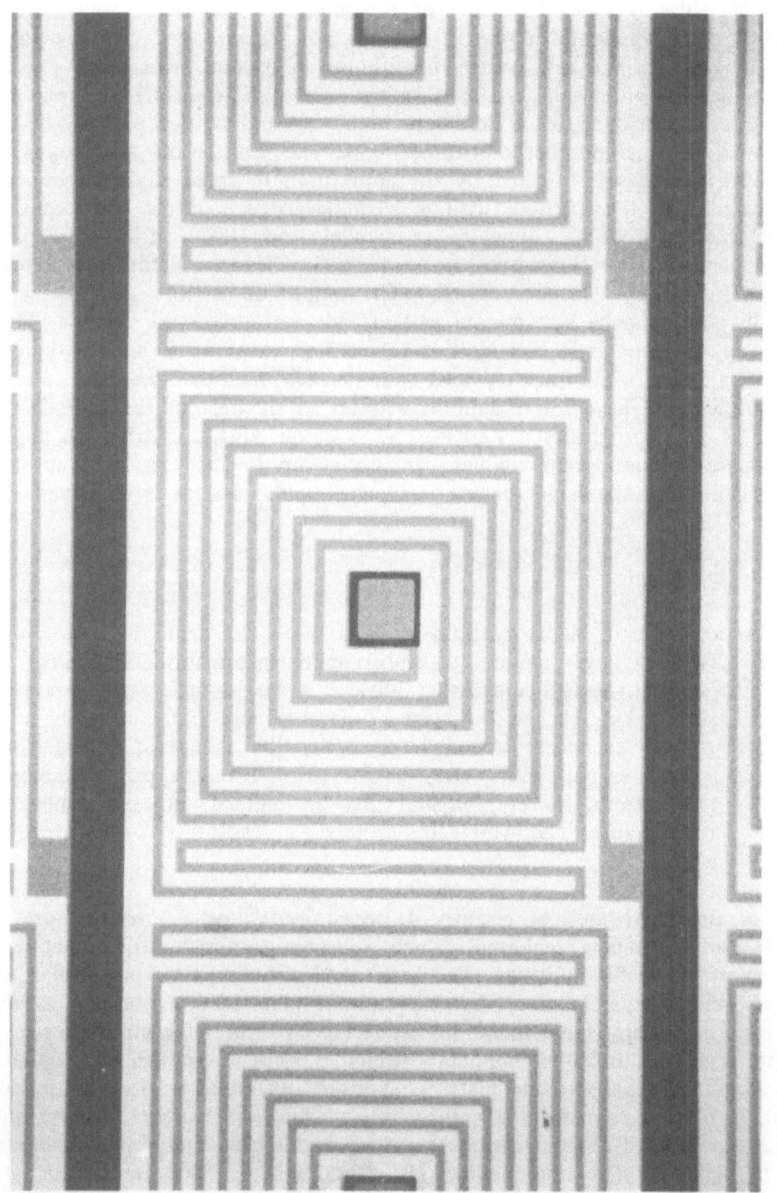

Fig. 12. Arrangement of the thin film resistor in the d.c. storage panel described by Smith[10].

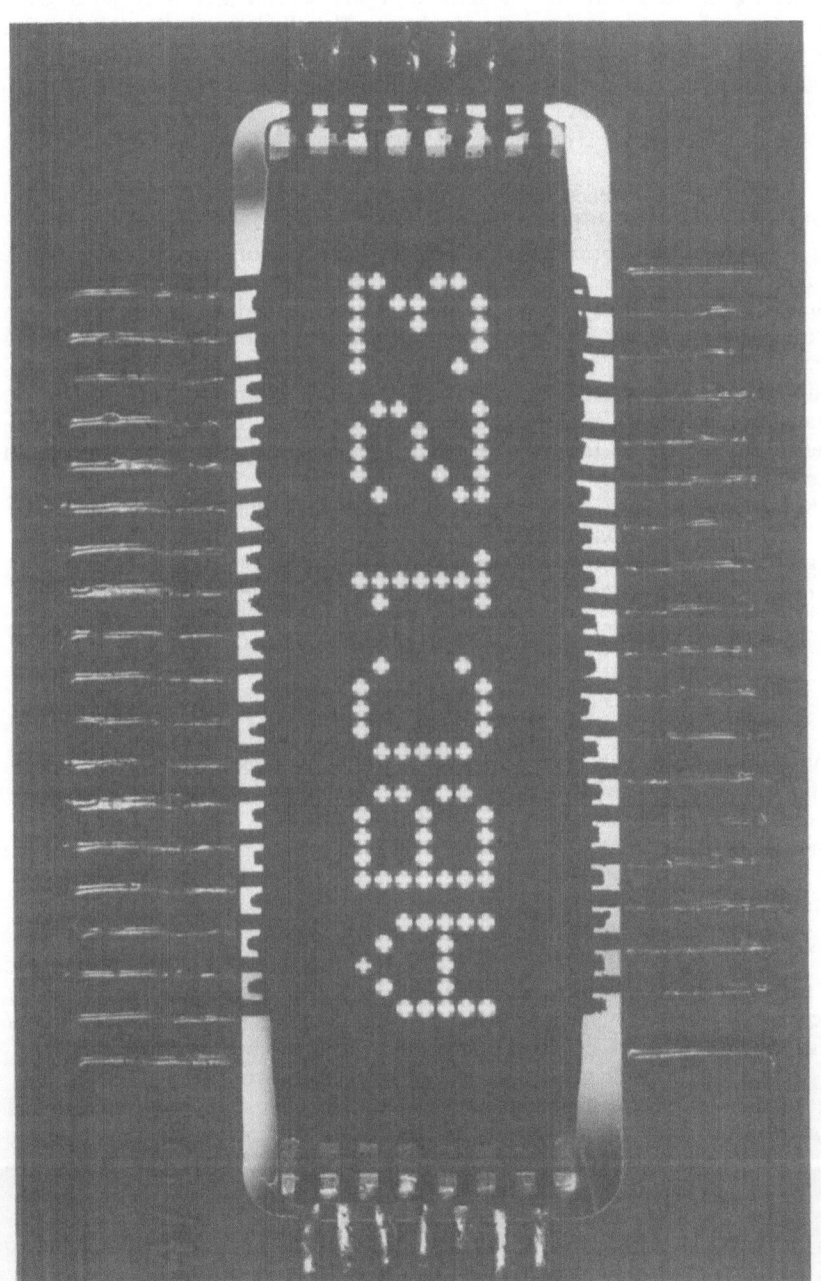

Fig. 13. A six-character storage panel using thin film resistors.

approach, the resistors are formed in the "land" between cross-bars. In this case, however, anode resistors are used, each being a fine spiral track of evaporated nichrome. The outer connection is to the cross-bar and the inner termination to an anode pad. Figure 12 illustrates the format. A thin silicon monoxide layer isolates the resistor and cross-bars from the discharge. The discharge is viewed through the resistors, light passing both through and between the spirals. The anode plate was used in conjunction with a moulded cathode plate. Similar brightnesses to the printed cathode-resistor panel have been achieved and Figure 13 shows a small experimental panel in operation. The cell pitch was 1 mm.

Both these approaches have problems of cell density and tolerances on characteristics. A completely different approach has recently been reported by Holtz (11) which overcomes the fabrication problems of high resistors in limited spaces. His solution is to apply pulses significantly greater than the cell firing voltage, but of short duration and low duty cycles, in place of the normal d.c. bias, using a standard "cyclic" type panel. Effective current limitation is obtained without the use of series resistor due, according to Holtz, to a combination of low duty factor and build-up time of the current during the pulse, which does not reach equilibrium. Only if the ionization is sufficiently high at the start of the short sustaining pulse will the cell turn on, and in fact only a cell which has previously been ignited will continue to re-ignite with each pulse. Thus, the panel has memory, the "off" cells failing to ignite and the "on" cells coming on during each pulse. The panel is addressed by pulses of much longer duration, illustrated in Figure 14. In terms of light output, the method lies between true d.c. storage and cyclic, with a fixed duty factor of 1/33. The principle is very similar to that used in the a.c. panel.

4.   A.C. PANELS

The idea of an a.c. panel was originated at Illinois University by Bitzer, Slottow and Willson (12). The first panels were similar in structure to the d.c. panels with three glass plates, except that the electrodes were coated on the outside surface of the outer plates, Figure 15. The external electrodes are capacitively coupled via the cell walls to the discharge walls, the equivalent circuit being also shown in Figure 15. When an a.c. signal is applied to the electrodes a reduced signal is applied across the gap.

Consider an a.c. signal applied across the electrodes, of such an amplitude V_f that a gas discharge breakdown can occur on each half cycle. The establishment of the discharge on, say, the positive half-cycle will result in the build-up of charge on the glass surfaces in front of the cathode which will set up a voltage V_{we} in opposition to the applied voltage. These wall charges have two effects (1) they can reduce the voltage during the half-cycle to a value below the extinction voltage and put out the discharge (which can happen in less than a microsecond) and (2) on reversal of polarity the wall charge voltage V_{we} adds to the applied voltage to allow breakdown at a lower applied potential (i.e. V_f-V_{we}). At any voltage between V_f and V_f-V_{we} the cell has a bistable characteristic; a cell initially ignited will re-ignite each half-cycle, but an off cell will never be ignited. Thus, one can consider breakdown and extinction potential analogous to the d.c. case, with, incidentally,

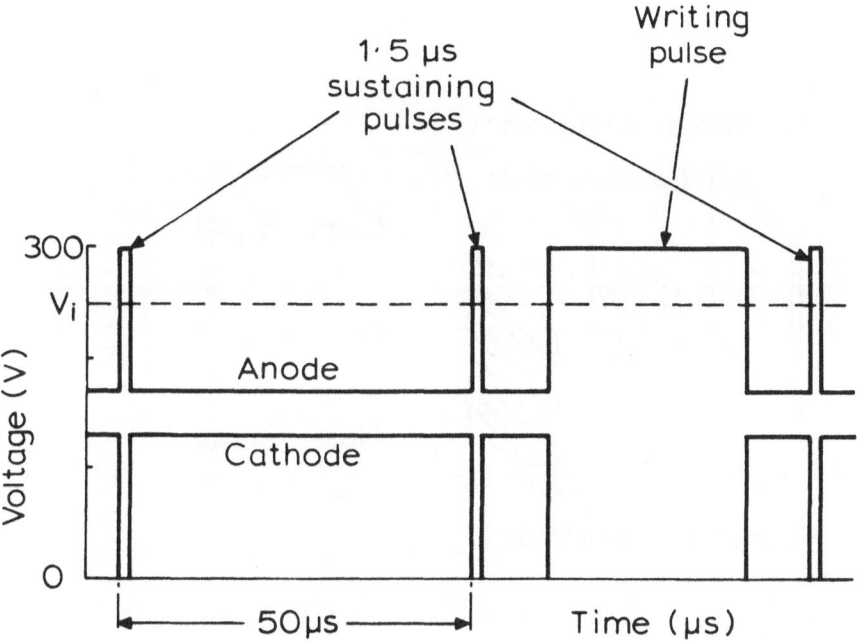

Fig. 14. Sustainer and address pulses for a d.c. panel driven in the pulse-storage mode.

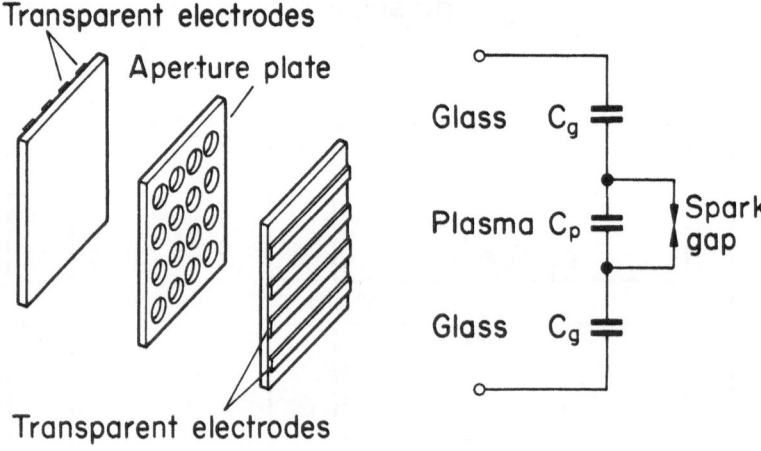

Fig. 15. Exploded view of an a.c. plasma panel, and the equivalent circuit.

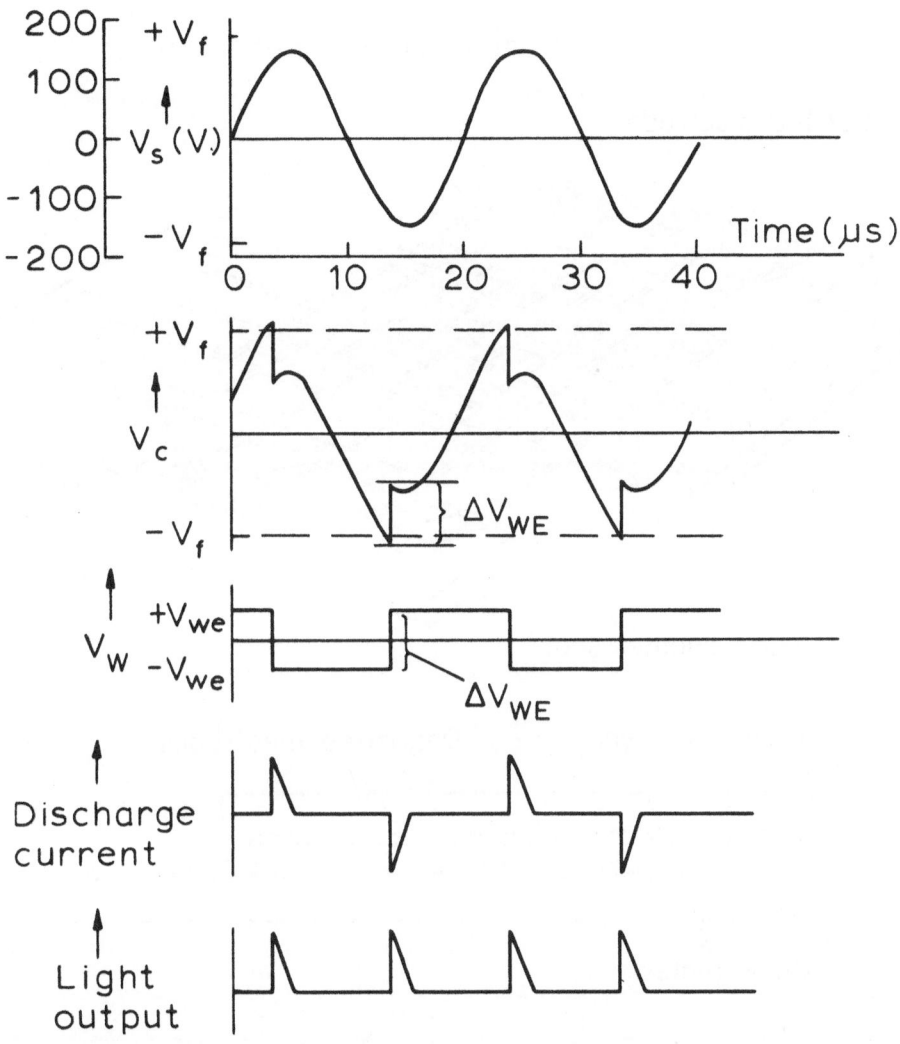

Fig. 16. The applied voltage V_s, tube voltage V_c and wall voltage V_w, waveforms for a plasma-cell together with current and light output.

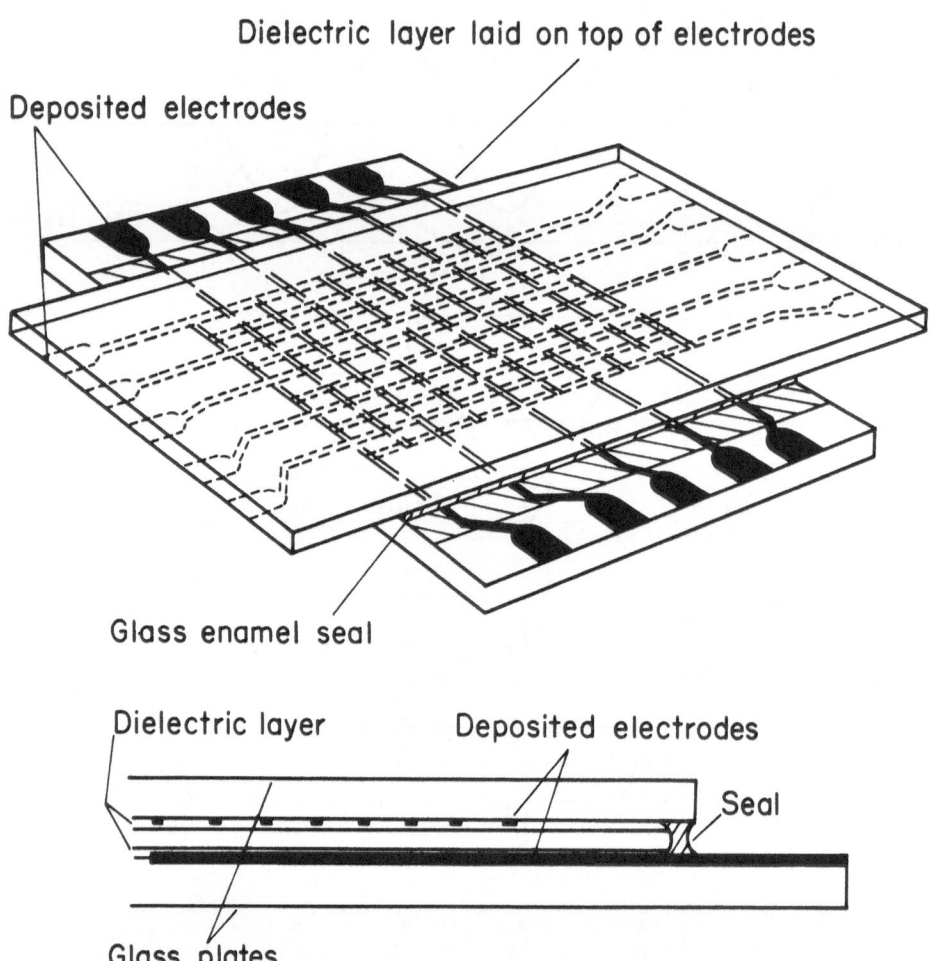

Fig. 17. Schematic diagram of the Owen-Illinois 'Digivue'[11] panel.

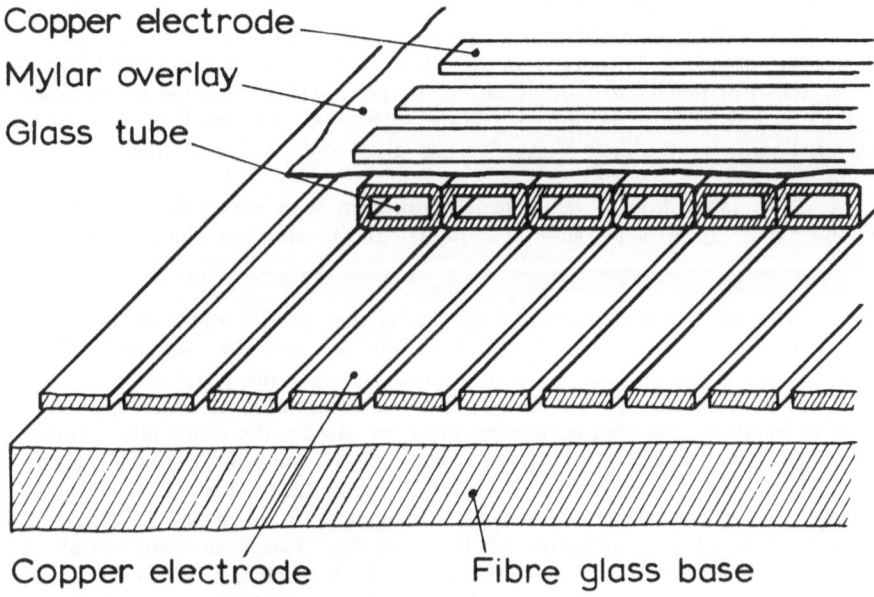

Fig. 18. Schematic diagram of the Control Data Corporation a.c. panel.

the same problems of spread and priming, and which is very similar to the pulse memory mode. This is illustrated in Figure 16 which shows the sustainer voltage waveform V_s and voltage across cell V_c and wall voltage V_w for a typical panel. In general, the wall charge is greatest when the discharge is short and of high intensity, and this to some extent depends on the gas pressure and composition.

Since the current pulse, also shown in Figure 16, is only on for a microsecond or so, the brightness will depend on frequency as well as the pulse amplitude. There is, however, a maximum frequency of about 100 kHz above which the panel will not function due mainly to cancellation of the wall charge by residual ions when the field is reversed. At this frequency the duty cycle is about one in a hundred. The amplitude of the current pulses will be determined by the signal voltage, and the capacitive impedance of the glass wall acts as the individual impedance for each cell. For maximum current at the lowest signal voltage the impedance should be as low as possible, inferring a thin glass wall. The small experimental panels used initially at Illinois University had glass walls 0.15 mm thick.

The fabrication of larger panels, however, with such thin glass walls would be extremely difficult, and therefore alternative structures have to be considered.

One of the most successful designs is that developed by Owen-Illinois Inc. described by Nolan (13). Essentially the electrodes are deposited on the inside surface of relatively thick glass plates (i.e. 6 mm thick) forming the panel walls, and subsequently coated by a thin glass dielectric film. In their design it was also found possible to eliminate the centre aperture plate by placing the outer plates close together. Thus the panel consists of two plates placed parallel so that the electrodes form a cross grid, the narrow space between being filled with gas and sealed round the edges with glass enamel. A schematic diagram of the panel is shown in Figure 17. Because of the simplicity of the structure, having no geometrical registration restrictions, a resolution up to 60 lines per inch is obtainable. Panels up to 512 x 512 lines are commercially available (14). An alternative approach has been developed by Control Data Corporation, described by Mayer and Bonin (15). In their design a thin capillary tube is used for each line of the panel. A cut away drawing of the panel is shown in Figure 18. The base consists of fibre glass with copper laminated to it, the lines being formed by photoetching. The tubes, 0.03 inch x 0.012 inch with wall thickness of about three mils, are laminated to the base using a clear epoxy. The top plate with the other electrodes is made from a copper-coated plastic sheet which is also laminated using the clear epoxy. The tubes are evacuated and then filled with the appropriate gas. The present resolution is 33 lines/inch and although a higher resolution looks difficult, they claim that double this resolution is feasible by having two electrodes along each tube. The main advantage is that the technique can be used for panels with large dimensions - a four foot long array has been constructed.

An interesting feature of the a.c. panel is the regularity of the light pulses from the cells giving no apparent jitter or ignition delay. This indicates a high degree of "self-priming", i.e. ionized particles formed during one discharge helps to initiate the following discharge. Also, the spreads in firing and extingui-

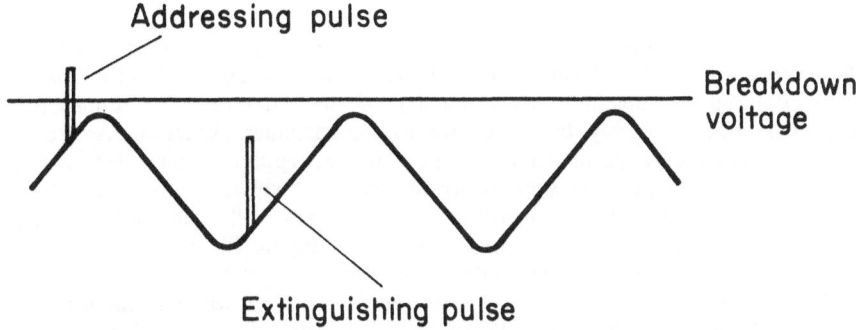

Fig. 19. Address pulses for sine wave sustainer voltage.

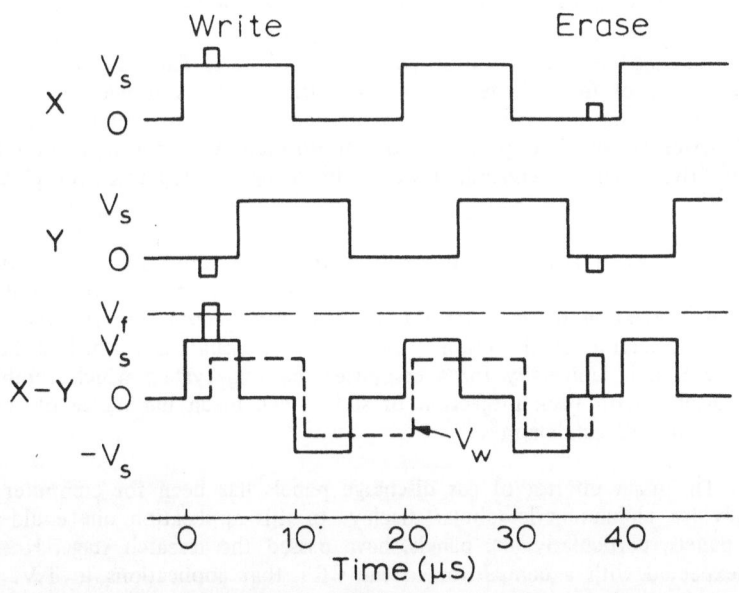

Fig. 20. Address pulses for stepped square wave sustainer voltage.

shing potentials appear to be rather lower than for the d.c. panel allowing reliability with a rather narrower operating margin.

To change the state of an a.c. cell it is necessary to alter its wall voltage, i.e. from zero for the "off" cell to a finite value V_w for an "on" cell. This is normally achieved by superimposing addressing of the order of 2 μsec duration on the sustainer voltage suitably timed relative to the sustainer waveform. As for the d.c. panel, ignition of a particular cell is obtained by applying co-incident pulses to the appropriate electrodes, the sum of which takes the voltage above the ignition value, whereas a single pulse is insufficient to ignite any cell. The address pulses for a sine-wave sustainer waveform are illustrated in Figure 19. The extinction pulse is applied before the positive cycle, and essentially produces a spurious discharge just sufficient to reduce the wall charge to zero, so that it will not fire on the following positive cycle. The principle can be seen more clearly in Figure 20, where the address pulses are shown for a square-wave sustainer waveform, together with the wall voltage values. The square-wave sustainer is preferred and has now superseded the sine-wave drive for a.c. panels. It has the advantage of being simpler to obtain circuit-wise and allows a greater tolerance on voltages. Also, the timing of the address pulses is less critical (the only criterion being that they must occur some time during the appropriate plateau period).

Because of the inherent memory, the panels can be randomly addressed and erased. The basic drive circuit is shown in Figure 21. The addressing signals could be applied by a logic tree system as illustrated in Figure 22 for a 4 x 4 array. Binary input words from the computer giving the X and Y location of the cell to be addressed or erased are stored in the registers. The output of the registers controls the state of the switches in the logic tree array. A register of n bits will control 2^n - 1 switches to address 2^n electrodes. Each row or column will have an associated driver D, which is pulsed at the appropriate time for write or erase. The number of drivers can be reduced, however, by using a submatrix multiplexing system.

The system is more complex than the cyclic drive for d.c. panels, but much more versatile, and applicable to larger panels. It is particularly suitable for graphics. An illustration of an Owen-Illinois panel being used for graphics is shown in figure 23. A feature of the Owen's panel is its transparency, which is being exploited by Illinois University for a computer teaching system which combines the panel display with back projection of slides. The mean luminance of the display was reported at 170 candela/m^2.

The main interest of gas discharge panels has been for computer terminals, mainly for alphanumeric tabular display. In this application one could claim that discharge panels, particularly a.c. panels, have passed the research stage. However, it is not unexpected with a competitor to the CRT that applications to T.V. should also be considered, and in this area such displays are still in research. The two major issues are the variation in brightness, i.e. half tones, and the production of full colour. Work on both these subjects is being carried out in the United States and elsewhere using a.c. and d.c. operated panels.

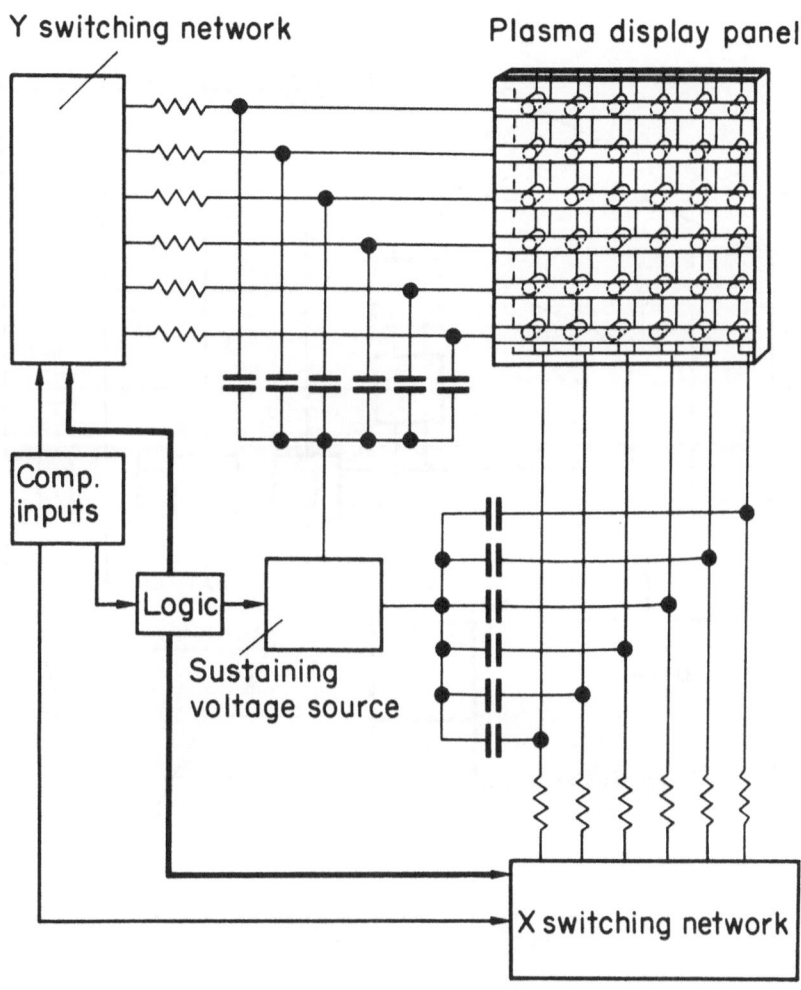

Fig. 21. Block diagram of the basic drive circuit for an a.c. panel.

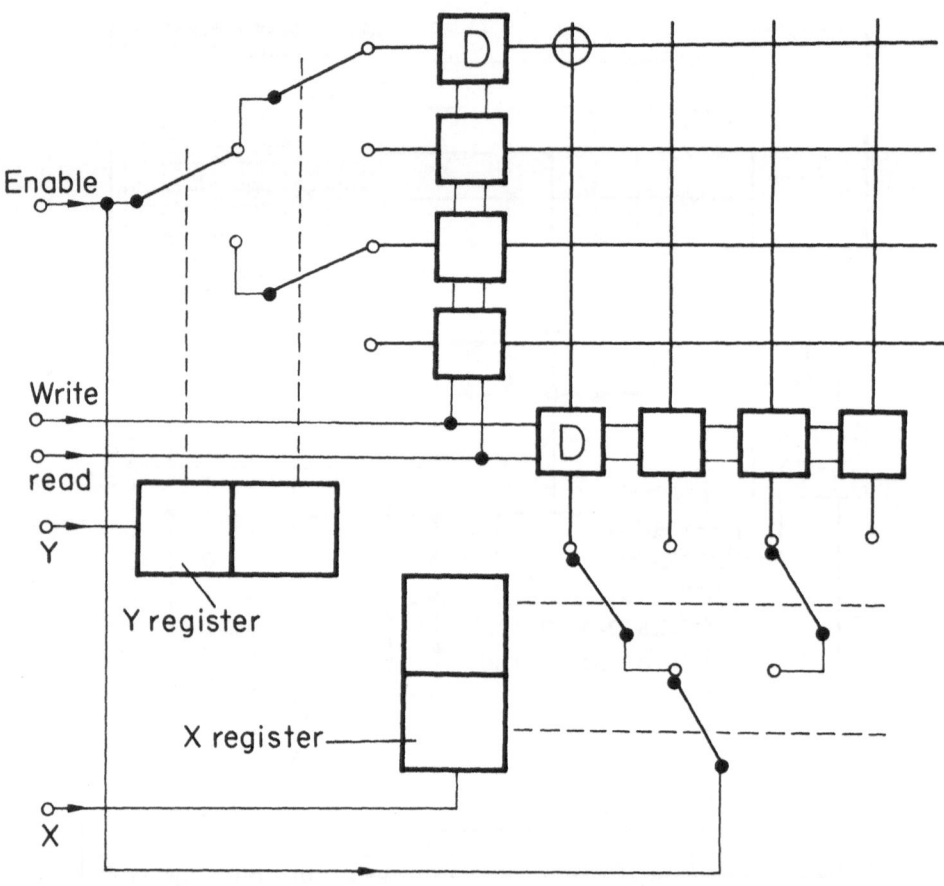

Fig. 22. Block diagram of address system for random access.

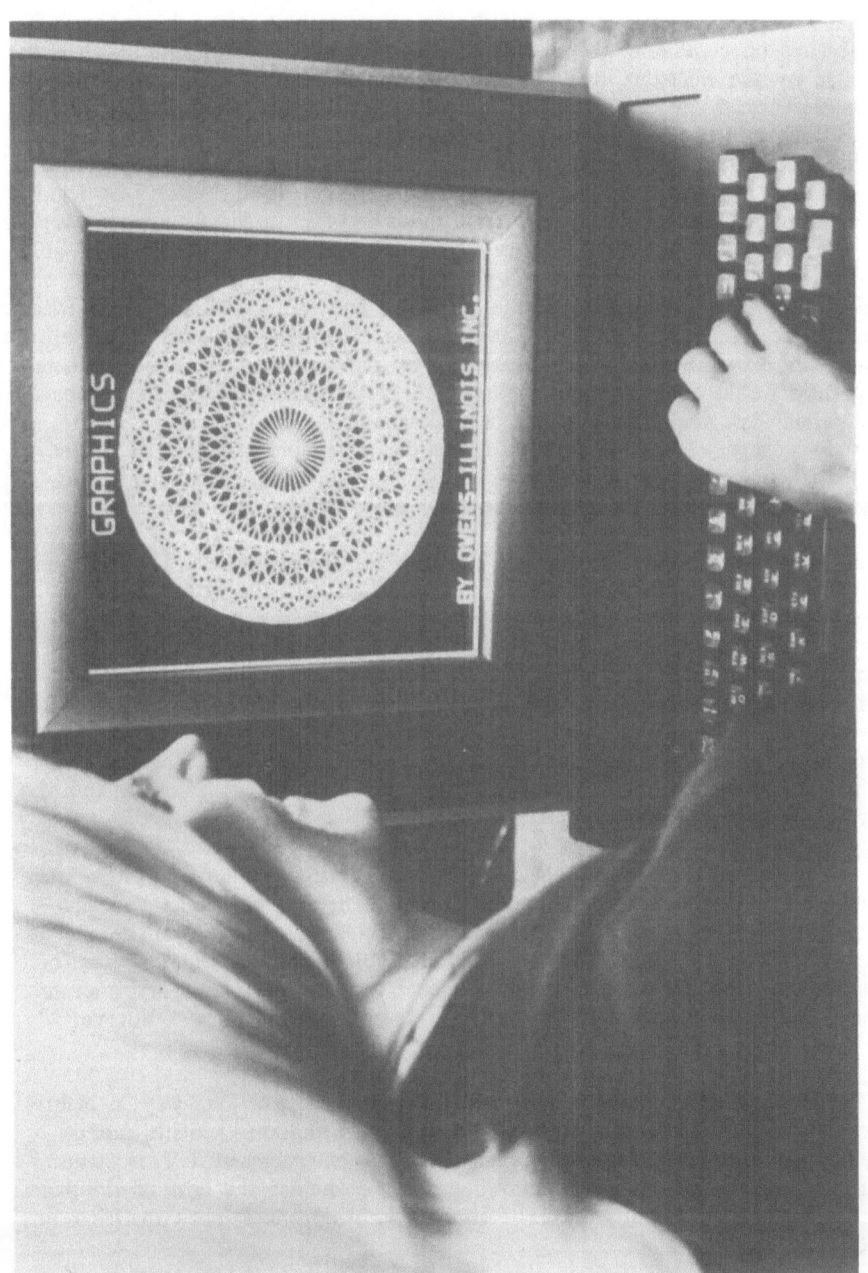

Fig. 23. Graphic display on an 8½ inch Owen Illinois panel having 512 × 512 lines with a resolution of 60 lines/inch.

Since it would be difficult if not impossible to consider cells containing different gases in the same panel to produce different colours, and in any case there are no gases which give efficient output in the green or blue, the achievement of colour depends on combining the discharge with phosphors. The phosphor can either be activated by the electrons, cathodo-luminescence, or be ultra-violet stimulated, photoluminescence. Some early work by Stredde (16) suggested that the former could give higher intensities, however, most of the experiments have been carried out with photoluminescent phosphors. In most cases standard panels have been used. The gas mixture is selected to give the maximum emission in the ultraviolet region with a minimum in the visible. Xenon or krypton is often used, or mercury vapour, and the phosphors are deposited as dots on the window in front of the cells, (17) or in the case of a d.c. panel on the walls of the individual cells (18). Problems arise due to the background glow of the discharge, and also cross coupling between cells. In the a.c. panel the deposited phosphor can affect the electrical characteristics. In most of the experiments efficiencies are lower than for the pure neon discharge and brightness would be insufficient for a colour T.V. display of 625 lines. One approach to increase the efficiency has been described by van Houten et al. (3) whereby the positive column is used to excite the phosphors. An analysis of the expected efficiency has been given by van Gelder and Mattheij (19) which unfortunately still falls short of the requirements for colour T.V.

Half tones present an especial problem for storage panels. The brightness can only be varied by varying the duty ratio, which is difficult with a common sustainer waveform whereby the discharge can be on either every cycle or up to every third cycle, depending on when it was addressed, de Jule and Chodil (20) were able to demonstrate four intensity levels including "off", but it is unlikely that this could be extended to a wider range. There is also the possibility of putting several panels behind each other with neutral density filters in between (21), or having such a high resolution that each picture element comprises several discharge points. In general these techniques have not been very successful, and the latter would probably be uneconomic.

Cyclic address panels are easier to adapt. In a line dumped system each dot can be either current or time modulated during the line period to give an acceptable intensity variation. de Boer (5) demonstrated the principle in 1968 and more recently Chen and Fukui (22) and Chodil, de Jule and Markin (23) have built systems based on the Burroughs "Self-scan" panel to show part of a T.V. picture. However, a cyclic panel with 625 lines to give acceptable brightness is not yet feasible (see section 2).

In summing up. the gas discharge panel has come a long way in eight years to the point where panels of over a quarter of a million elements can be produced in 400 cm^2 for computer applications. The extension to T.V. is some way off, and there is still the important question of whether the cost of the panel and its associated drive circuits makes it a really viable product in competition to the C.R.T.

REFERENCES

(1)   G. E. Holtz, Proc. Soc. Inf. Disp. 13, 2 (1972).

(2)   Electronic News, p. 5, 26th July, (1965).

(3)   S. van Houten, R. N. Jackson and G. F. Weston. Proc. Soc. Inf. Disp. 13, 43 (1972).

(4)   R. F. Hall, K. E. Johnson and G. T. Sharpless, I.E.E. Conference publication, No. 80, 91 (1971).

(5)   Th. J. de Boer, Proc. 9th Nat. Symp on Inf. Disp. 193 (1968).

(6)   R. N. Jackson and K. E. Johnson, IEEE Trans Electron Device E.D. 18, 316 (1971).

(7)   W. J. Harman, Electronics 43, 20 (March 2nd, 1970).

(8)   Th. J. de Boer, Conference Digest 1969 IEEE Int. Electron Device Meeting 52 (1969).

(9)   F. Walters, I.E.E. Conference publication no. 80, 7 (1971).

(10)   J. Smith, To be published in IEEE Trans. Electron Device E.D. 20, Nov. (1973).

(11)   G. E. Holtz, Digest of Papers, 1972, Symp. of Soc. Inf. Disp. 36, (1972).

(12)   D. L. Bitzer, H. G. Slottow, and R. H. Willson, Co-ord. Science Lab., Quart. Prog. Report No. 31. University of Illinois (1964), also Proc. 1966 Joint Computer Conference. Spartan p. 541 (1966).

(13)   J. F. Nolan, Conference digest, IEEE Int. Electron. Device. Meeting (1969).

(14)   H. J. Hoehn and R. A. Martel, IEEE Trans. Electron Device E.D. 18, 659 (1971).

(15)   W. N. Mayer and R. V. Bonin, IEEE Conf. record of 1972 Conf. on Display Devices, 15 (1972).

(16)   E. Stredde, Co-ord Science Lab. report No. R370, University of Illinois (1967).

(17)   F. H. Brown and M. T. Zayac, Proc. Soc. Inf. Disp. 13, 52 (1972).

(18)   J. Forman, Proc. Soc. Inf. Disp. 13, 14 (1972).

(19)   Z. van Gelder and M. M. M. P. Mattheij, Proc. IEEE 61 1019, (1973).

(20)   M. C. de Jule and G. J. Chodil, Proc. 1971 S.I.D. Conf. Disp. 102, (1971).

(21)   D. T. Ngo, Proc. Soc. Inf. Disp. $\underline{13}$, 21 (1972).

(22)   Y. S. Chen & H. Fukui, IEEE Conf. record of 1972 Conf. on Display
          Devices 70 (1972).

(23)   G. J. Chodil, M. C. de Jule and J. Markin, IEEE Conf. record of 1972
          Conf. on Display Devices 77 (1972).

PRESENT STATE OF THE DIGITAL LASER BEAM DEFLECTION TECHNIQUE FOR ALPHANUMERIC AND GRAPHIC DISPLAYS

U. J. Schmidt

Philips Forschungslaboratorium Hamburg (West Germany)

INTRODUCTION

Some ten years ago lasers were considered excellent candidates for bright displays. In the intervening years the laser display field did not seem to move much and quite a few contenders have appeared since, as manifested in the large number of devices for display applications described at this seminar.

The two properties of lasers, high brightness and their optimal depth of color, made them seem most suitable for multichrome large screen displays. However, in spite of the fact that the laser is indeed many orders of magnitude brighter than thermal light sources, the actual brightness of standard commercial laser systems did not quite reach the expected power level necessary for large screen displays. For screens of a few square meters in size and for a desired intensity of 1 W/m^2 the commercial systems have so far missed the target by about one order of magnitude. Detrimental to an application is also the low efficiency of the often used ion laser with $\eta \leqslant 0.005$ %. This paper will report on one particular type of large screen display where this barrier has been overcome by the development of a deflection technique that in effect compensates for the low laser powers available.

The above figure of 1 W/m^2 at the display screen is only valid for line scanned pictures where continuous intensity (amplitude) modulation of the light beam is required. There is, however, the class of alphanumeric and graphic displays where information has to be displayed only in isolated locations of the screen. No continuous scanning would be required here and, consequently, no blanking of the beam if a truly random access beam deflection method were on hand. In contrast to some scanning methods "random access" in this context means that the access time to any one position of the screen is short as compared to the dwell time of the beam. In this case, no blanking would be necessary in that the beam would be deflected to only those positions where information had to be displayed. The savings

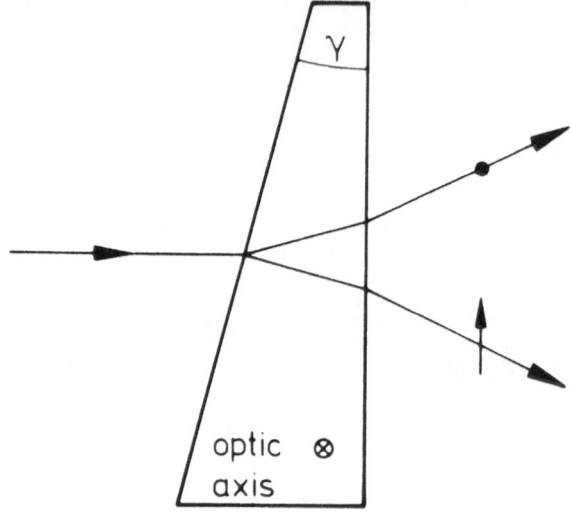

Fig. 1. Digital deflection by birefringent prism.

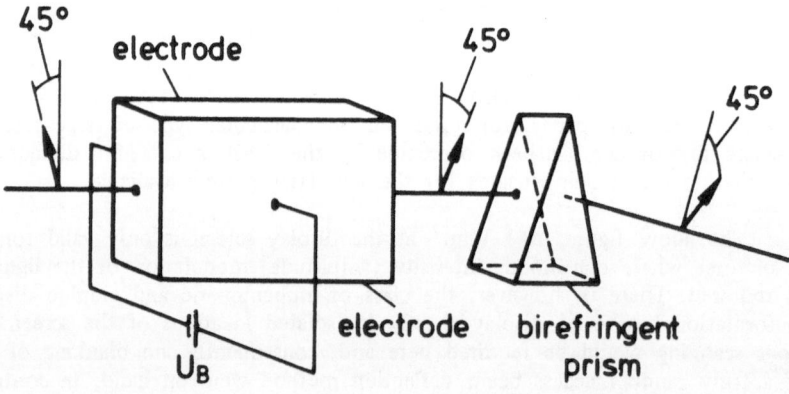

Fig. 2. Digital deflection stage consisting of Kerr cell and birefringent prism.

in light power would be given by the ratio of the total screen area to the information-covered area. For many alphanumeric and graphic displays this ratio is considerably larger than ten, bringing the light power requirements down to the commercially available laser powers.

THE DIGITAL DEFLECTION TECHNIQUE

Fig. 1 depicts the well known phenomenon of birefringence being the basis of the deflection technique to be described (1, 2, 3). A beam of light may be deflected to either one of two directions by modulating the linear polarization state of the beam between one oriented parallel and one oriented perpendicular to the optic axis of the birefringent prism. This optic axis is in a plane approximately splitting the refracting angle of the prism in half. The state of polarization may be controlled by a number of types of switches. Fig. 2 indicates a Kerr cell as used in this laboratory. The cell has to assume two states for zero degree and 90 degree rotation of the plane of polarization. To achieve this, only high voltage switches S_1 and S_2 are required as sketched out in Fig. 3 for charging the Kerr cell from a DC power supply and for discharging the cell which - in electronic terms - represents a capacitance. As the response time of nitrobenzene corresponds to the relaxation time of approximately $1/2 \times 10^{-10}$ sec, the switching time for changing the direction of the light beam depends essentially on the properties if the switches S_i.

The combination of a Kerr cell and a birefringent prism is called a deflection stage. A serial combination of a number n of such stages gives the possibility of addressing 2^n beam positions as shown for six stages in Fig. 4. A lens following the deflector will permit the conversion of the deflection directions into separate beam positions in the focal plane of this lens if the refracting angles of the prisms are properly chosen with respect to the angular aperture of the beam.

The properties of this deflection principle are as follows:
Each beam position is characterized by an n-dimensional binary word.
Apart from scattered light, no intermediate beam positions are possible.

Following the technical approach as schematically shown in Fig. 4. light losses in the visible region are low because of the high degree of transparency of nitrobenzene and calcite and because of the low reflection of 0.1 % at each interface nitrobenzene-calcite prism.

Since the refractive index of nitrobenzene is nearly the arithmetic mean of the refractive indices of calcite, the two alternate deflection directions subtend in each stage an angle of the same amount but of opposite sign at the incident beam direction. This keeps the system axis straight without further measures.

The Kerr cells are driven in parallel, each by its own switch combination S_1, S_2 (Fig. 3). This makes the deflector a random access device: Any position of the pattern is reached by the beam within the same time, independent of the previous position.

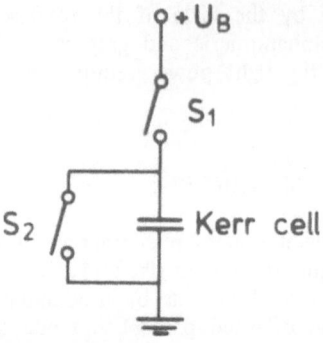

Fig. 3. Electronic circuit for the 2-state-operation of a Kerr cell.

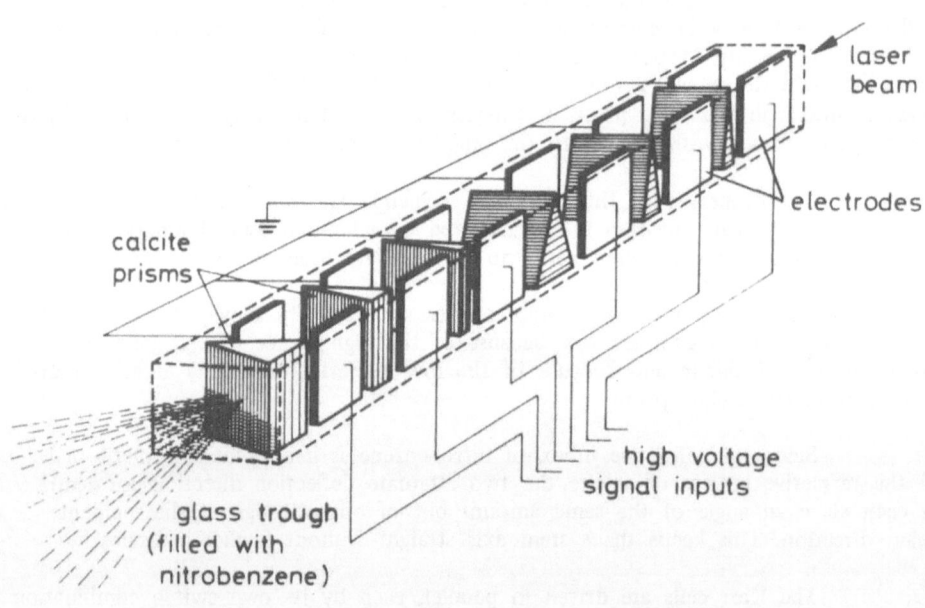

Fig. 4. Schematic diagram of a 6-stage digital beam deflector.

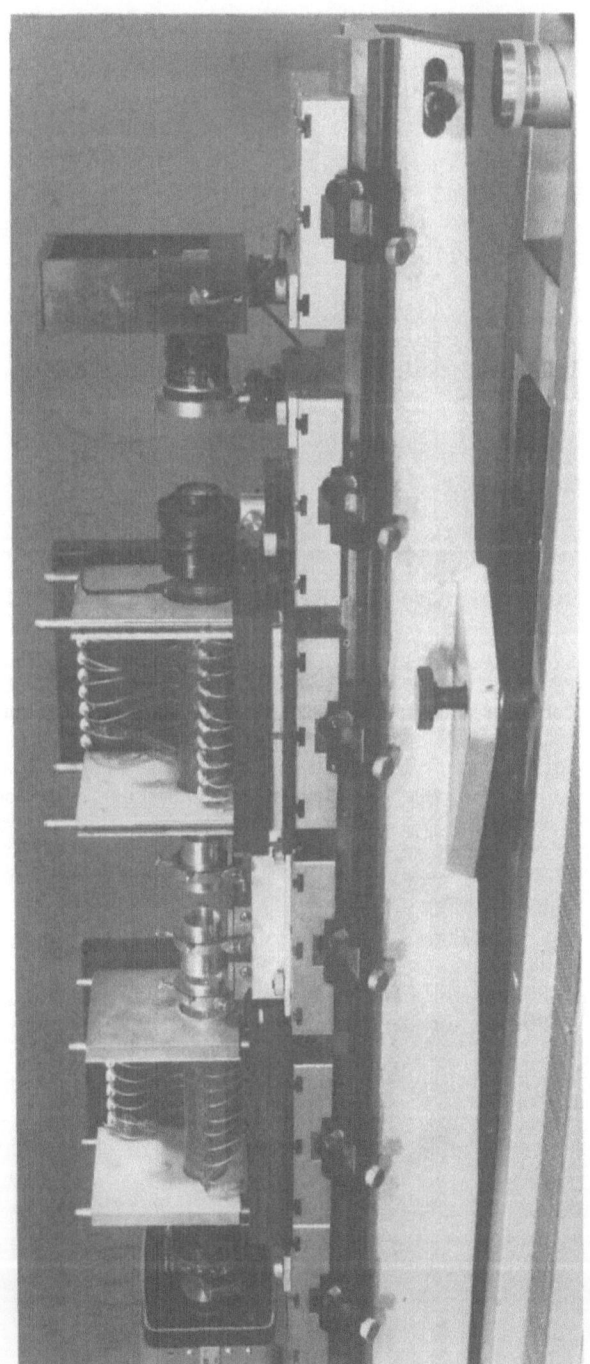

Fig. 5. Photograph of 20-stage digital laser beam deflector.

Table 1. FEATURES OF EXPERIMENTAL 20-STAGE DIGITAL DEFLECTOR

Resolution: 1024 x 1024 beam positions, neighbouring positions half overlapping

Transmission: 80 %

Scan angle of deflected beam: ± 4.36°

Exit beam diameter: 2.8 mm

20 % increase of zero spatial mode Gaussian beam halfwidth

Length of 10-stage deflector: 22 cm

Length of 20-stage deflector including polarization filtering stage: 85 cm

Accuracy of prism adjustment: ± 10 minutes of arc in the plane perpendicular to
                                the system axis, corresponding to 0.5 beam positions

Accuracy of electrode width adjustment: ± 0.02 mm

Maximum scanning rate: 500,000 s^{-1} for first ten stages
                        60,000 s^{-1} for the other stages

Random access time: 300 ns for first ten stages
                    900 ns for the other stages

Signal to background ratio: better than 100 : 1

Table 2. DESIGN DATA OF 20-STAGE DEFLECTOR

Stage no.	Prism angle d stage	Prism angle h stage	Optical aperture d(mm) × h(mm)	Electrode separation (mm)	Voltage (for λ = 520.8 nm) Bias (kV)	Voltage (for λ = 520.8 nm) Switching (kV)
1		6'	1.4 × 1.4	1.6	5.3	2.1
2		12'	1.4 × 1.4	1.6	5.3	2.1
3		24'	1.4 × 1.4	1.6	5.3	2.1
4		48'	1.4 × 1.4	1.6	5.3	2.1
5		1°36'	1.4 × 1.5	1.6	5.3	2.1
6	6'		1.4 × 1.6	1.6	5.3	2.1
7	12'		1.4 × 1.7	1.6	5.3	2.1
8	24'		1.4 × 1.8	1.6	5.3	2.1
9	48'		1.4 × 1.9	1.6	5.3	2.1
10	1°36'		1.5 × 2.0	1.6	5.3	2.1
11		1°36'	3.0 × 4.0	4.5	14	6.5
12		3°11'50"	3.0 × 4.1	4.5	14	6.5
13		6°22'20"	3.0 × 4.2	4.5	14	6.5
14	1°36'		3.1 × 4.6	4.5	14	6.5
15	3°11'50"		3.1 × 4.9	4.5	14	6.5
16	6°22'20"		3.3 × 5.3	4.5	14	6.5
17		12°35'15"	3.6 × 5.6	4.5	14	6.5
18		−25°11'30"	4.0 × 6.4	4.5	14	6.5
19[a]	12°35'15"		4.7 × 7.0	6.0	20	8.8
20[a]	−25°11'30"		5.5 × 8.6	6.0	20	8.8

[a] The two last prisms are inverted, bringing a good temperature stabilization of the deflection pattern.[7]

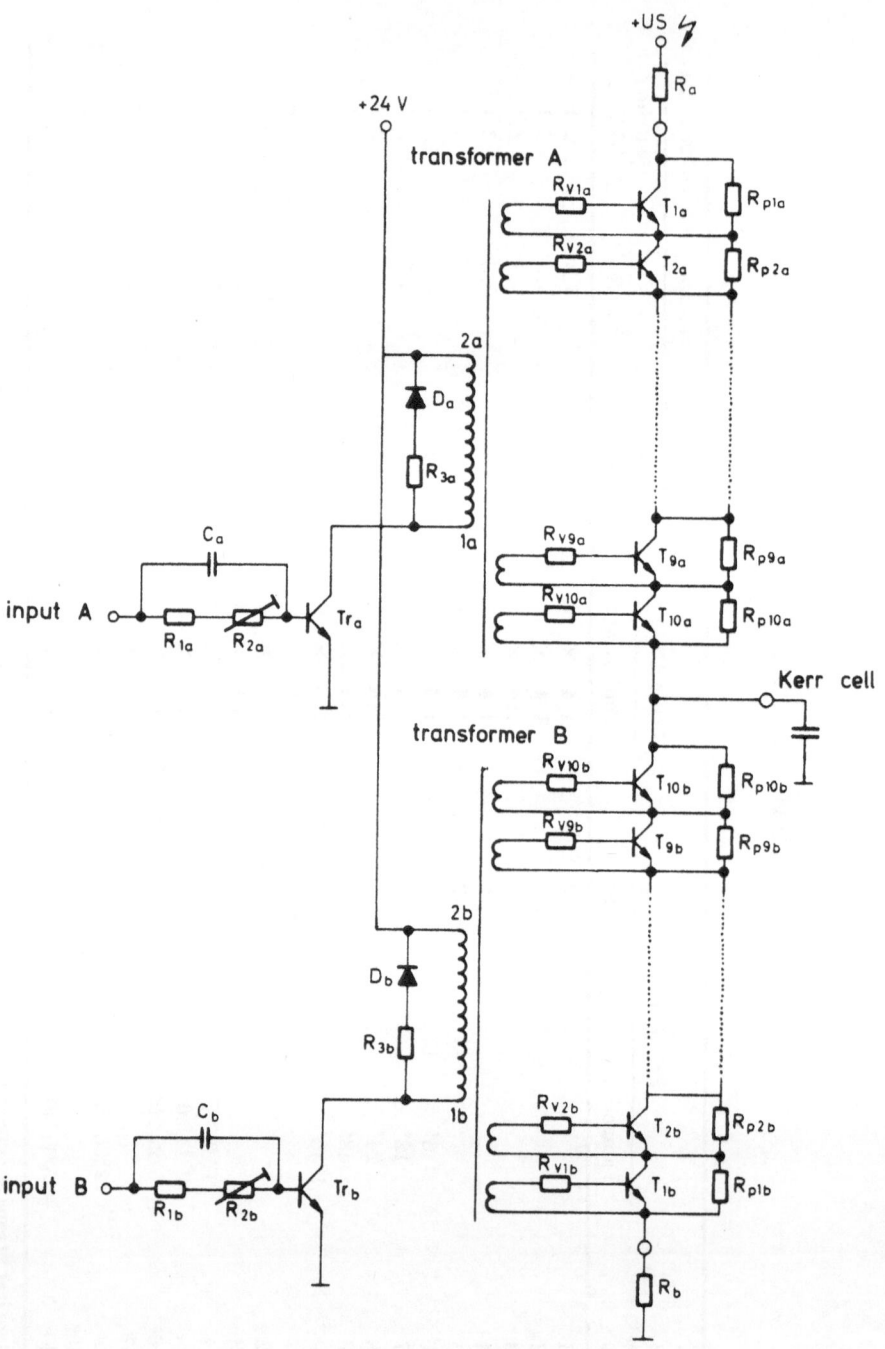

Fig. 6. Detail diagram of electronic high voltage switch.

EXPERIMENTAL RESULTS

Along the technical approach as sketched in Fig. 4 digital deflectors up to 20 stages have been designed and built (4), Fig. 5. The detailed considerations, regarding the choice of materials, construction techniques and design optimization are dependent on a relatively large number of physical, mechanical, chemical and electronics aspects. To give an account of these goes beyond the scope of this paper and reference should be made to the literature (5,6).

In the context of this paper it may suffice to give the more relevant data of the 20-stage deflector. They are listed in Table 1 and Table 2. Resolution, speed and light transmission are the most important parameters of a deflector. As will be discussed in the following sections, the reported values qualify for a number of display applications.

The experimental deflectors have been shown to be reliable devices. The first 20-stage deflector built has been operated for more than one and a half years with an operational record of 1700 hours. Also the high voltage switches have been found practical solutions. They have been realized by stacking 10 transistors each in series as shown in Fig. 6. To provide for an equal distribution of voltage across all transistors both a resistor and a capacitance have been connected in parallel with each transistor. The command pulse for each transistor base is derived from an isolated secondary of a common pulse transformer. The power of these pulses has to vary depending on the signal pattern (7). The voltages to be handled in a 20-stage deflector range between 6.3 kV for the first stages and 21 kV for the last two stages. In order to reduce these values use has been made of the quadratic characteristics of the electrooptic Kerr effect (Fig. 7). Application of a DC bias field for each cell for a constant 90 degree pre-rotation of the plane of polarization reduces the signal voltages by a factor of 0.4 down to 2.4 kV to 8.5 kV. A higher bias for further reduction of signal voltage is impeded by the danger of electric breakdown in the liquid.

A LASER DIGITAL DISPLAY

The digital light deflector as described above has been used as one of the basic elements in a computer controlled laser display projector (8). A block diagram of the system is shown in Fig. 8.

DESCRIPTION OF THE COMPUTER SOFTWARE

The software has been laid out for a real time display, i.e., the projection screen was not covered with a material showing afterglow. For this reason, the information had to be presented at a frame rate of about 60 s^{-1}. It proved to be practical to separate physically the information generating process computer from the display memory which stored each point in the display to be addressed by the laser beam as a 20-bit word. The computer software was subdivided into a function generator and the operational system. A basic diagram is shown in Fig. 9. The func-

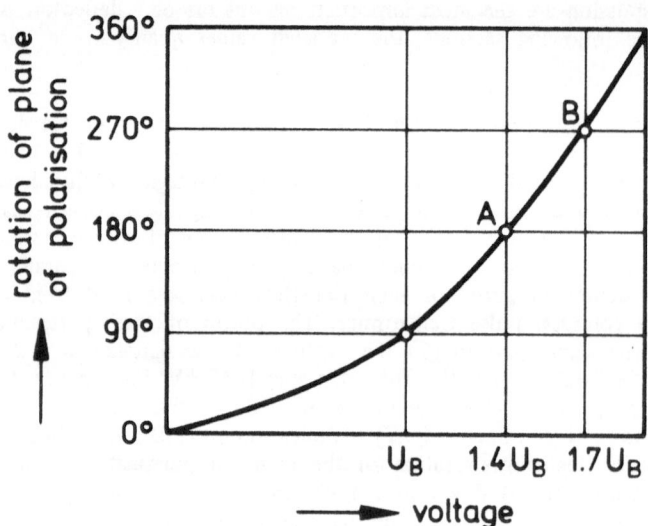

Fig. 7. Graphic representation of Kerr's Law.

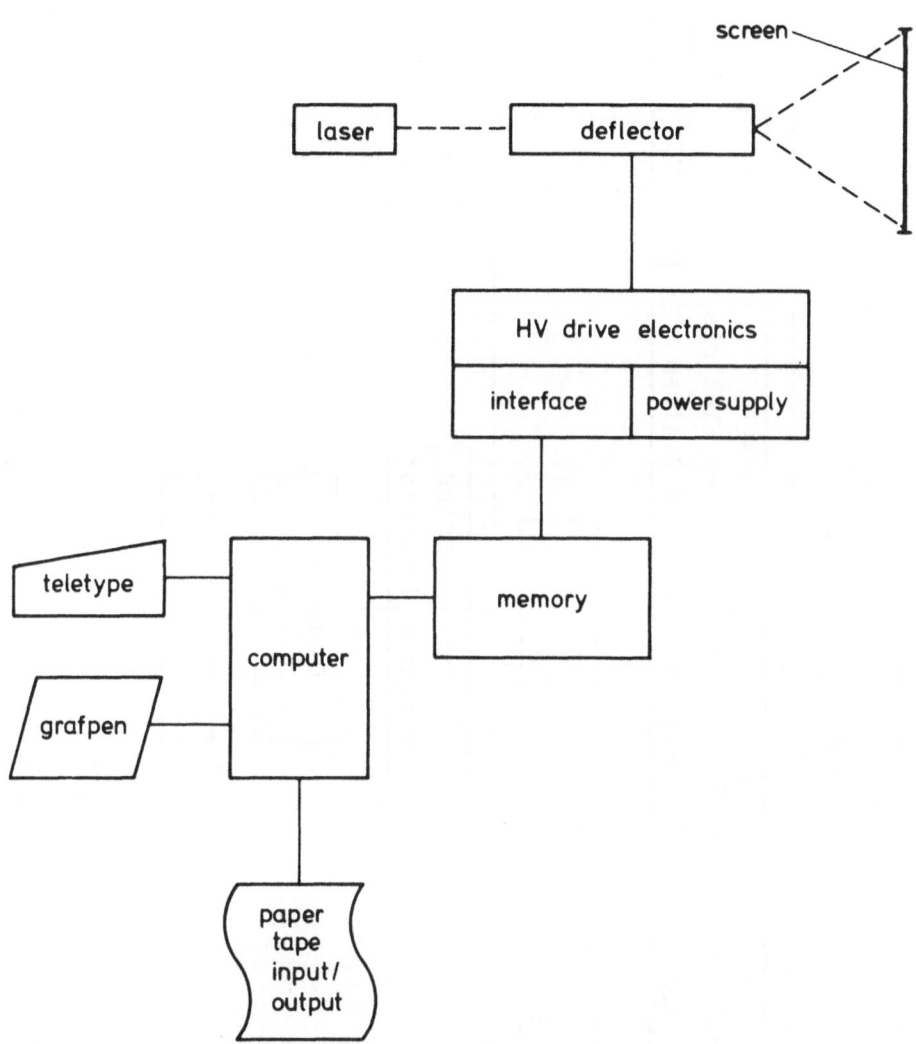

Fig. 8. Block diagram of laser display system.

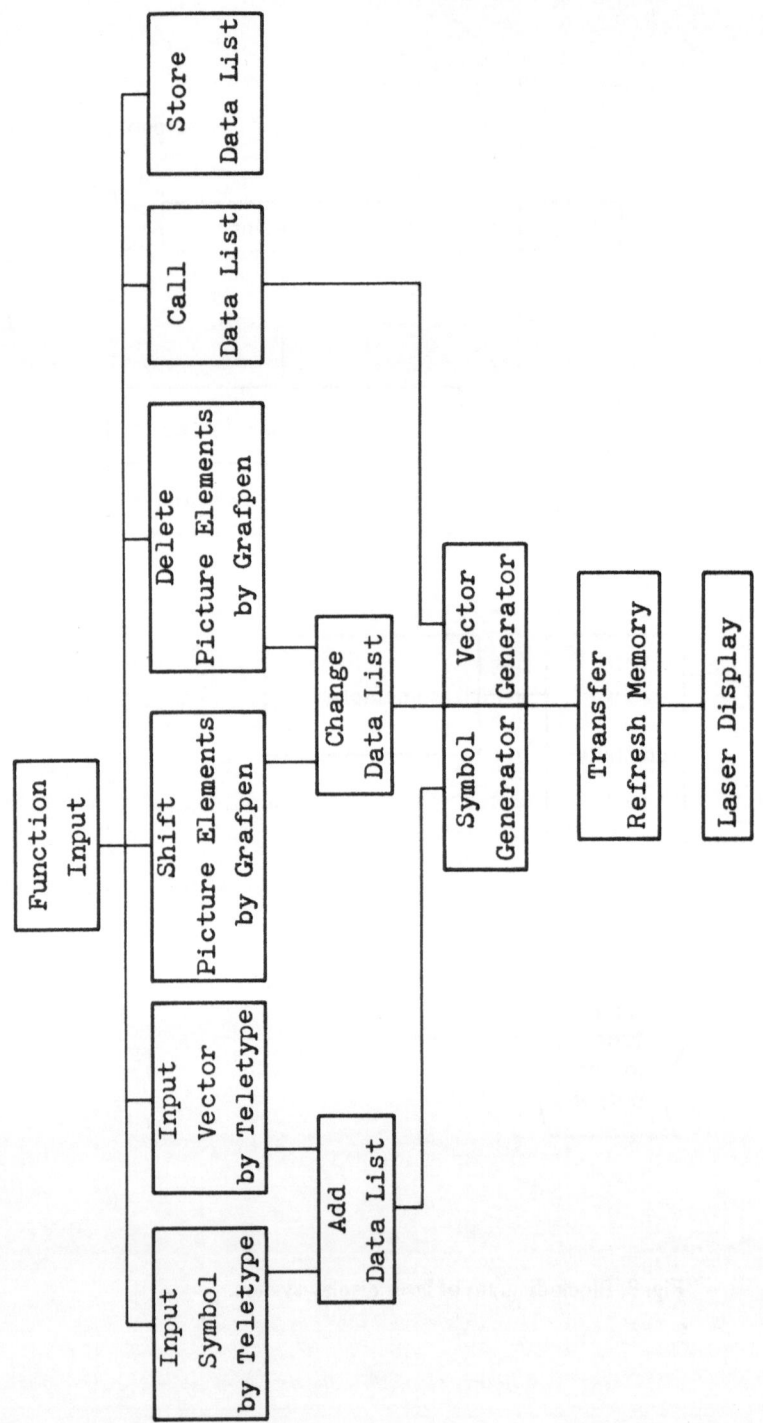

Fig. 9. Schematic diagram of software layout for display system.

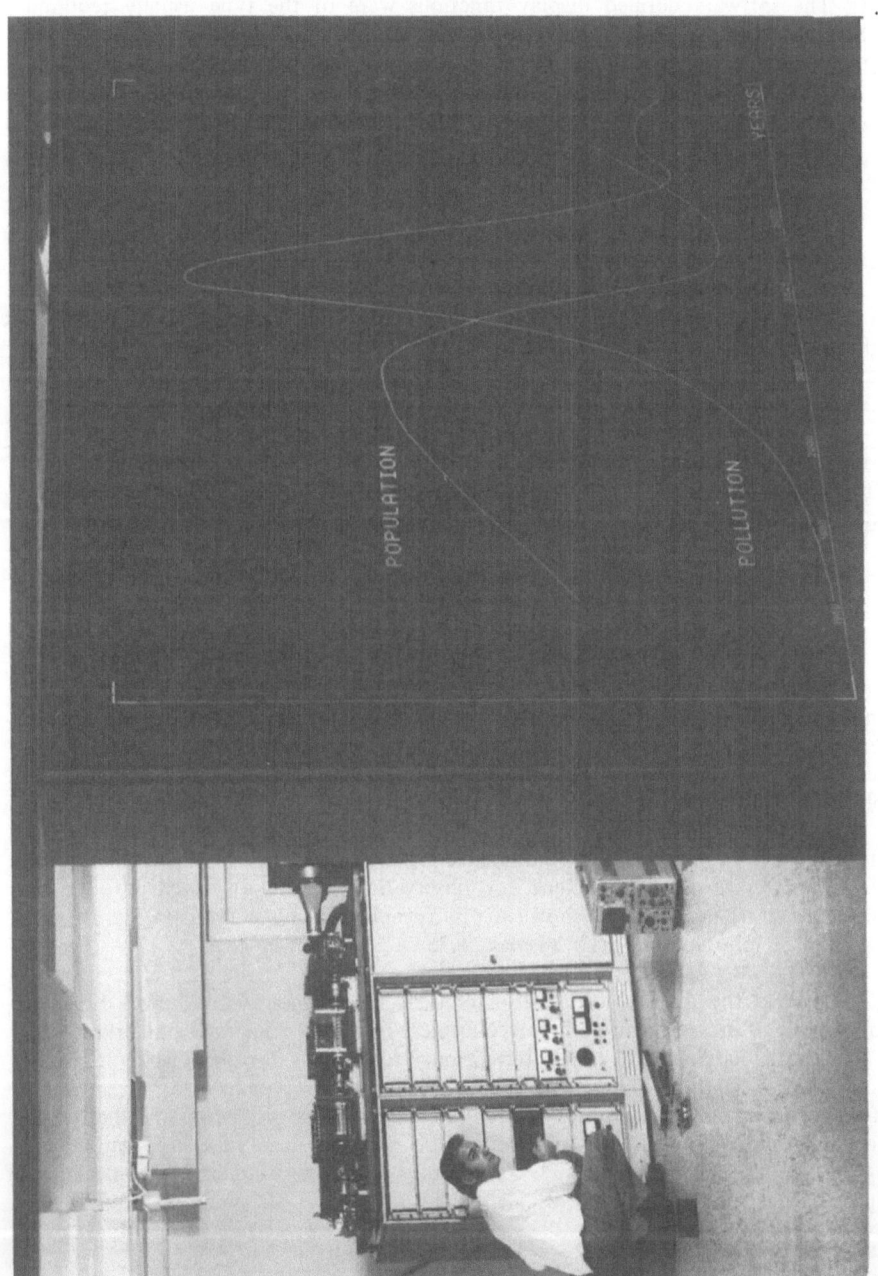

Fig. 10. Photograph of 20-stage laser display system for alphanumerics and graphics. Screen size: 2 × 2 m.

tion generator was programmed using vectors of 8 different orientations as basic elements. The software defined display functions were of the type usually required for alphanumeric and graphic displays for traffic control. The input means tested were paper tape, teletype, joy-stick and ultrasonic grafpen. The program was designed to permit on-line manipulation of the displayed information via these input devices. Of these various means the grafpen proved most attractive for positioning a symbol, vector or the like at a desired location on the screen via a travelling cross marker (cursor).

The capacity of the displayed information is, of course, determined by the position switching rate and the frame rate. For the values of 500,000 s^{-1} and 60 s^{-1}, respectively, a display may contain up to approximately 8000 points. Making use of the random access capability of the light deflector this capacity could be increased through proper software design. Analogous to the alternate line scan principle applied in TV each alternate neighboring beam position is addressed in a first full frame scan before the other half of beam positions is addressed. This way, the apparent frame rate is doubled. To pursue this method even further by scanning one frame in three or more go-arounds appears not satisfactory as the eye notices - at least from close distance - an apparent disturbing regular motion of points within each symbol. Random scan methods have not yet been tested.

Frame rates below 55 s^{-1} do not give a satisfactory flickerfree display. Higher rates are recommended whenever possible.

The size of the display is variable within wide limits and may range from microscopic dimensions to large screen pictures of several square meters. The brightness varies accordingly; of course, it depends also on a number of other parameters, e.g., the laser source power, the overall transmission of the system and the percentage of the frame area to be covered with information. To give an example, for many types of traffic control displays only about 3 percent of the frame area is to be illuminated. Using a laser of 500 mW output and having a system of 30% transmission, the illuminance at a screen of 2 X 2 m is approximately 2000 lux, which is sufficient for observation of the displayed information in a normally lit room. Fig. 10 shows a photograph of such a system. Fig. 11 gives a close-up photograph of a display 1.25 X 1.25 m in size.

Often, the display information consists of a time-variable and a time-constant fraction. In such a case, it is economical regarding scan rate and brightness to separate both fractions by using conventional techniques for the display of the time-constant information: slide projection as well as map type projection screens were found usable, for both front view and rear view type projection methods. Even ordinary topographical wall maps could be used for rear view observation although at an expense in brightness of about one order of magnitude. Because of the high brightness level the symbols are clearly visible on all types of overlays, even on multicolor maps.

Converting the single laser wavelength laser projector to a two or multi-color system is a nearly straightforward affair if a separate deflection system is used for every additional color. Modifications arise mainly in the design of the

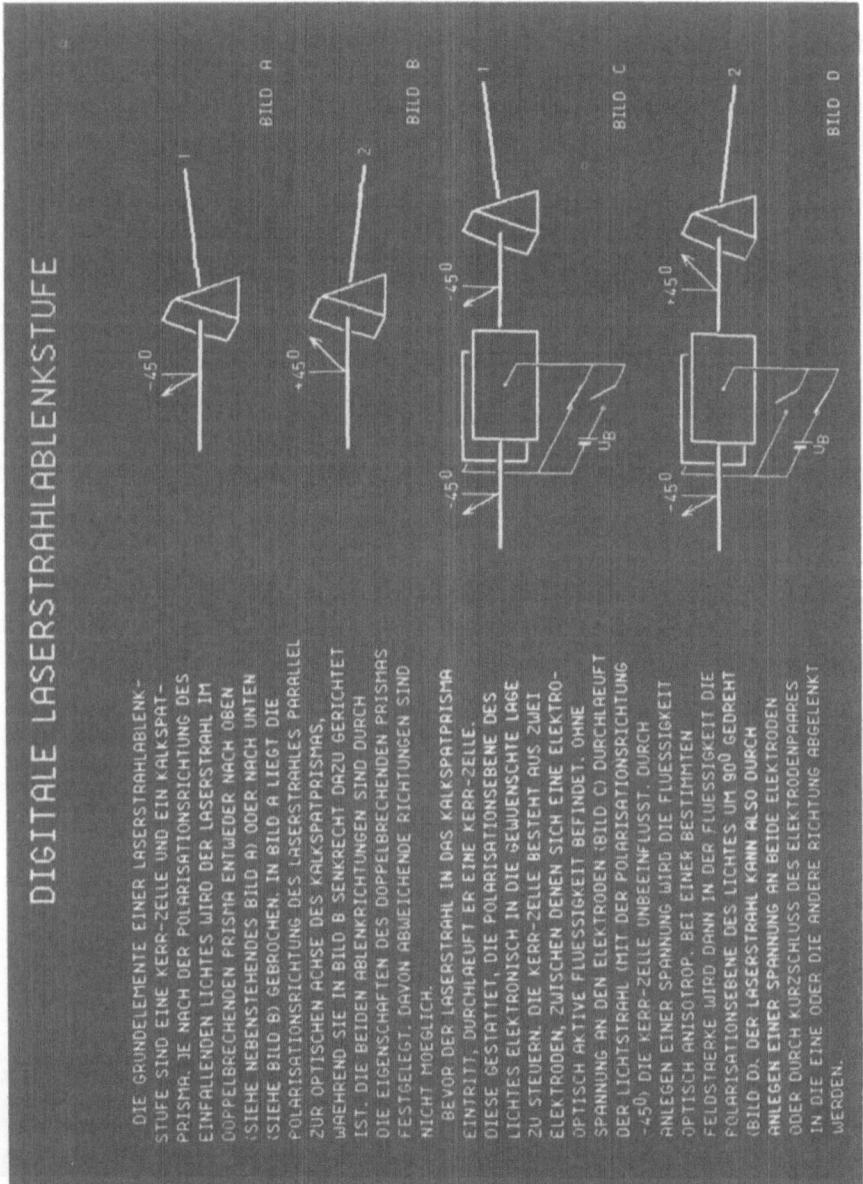

Fig. 11. Photograph of a display 1.25 × 1.25 m in size.

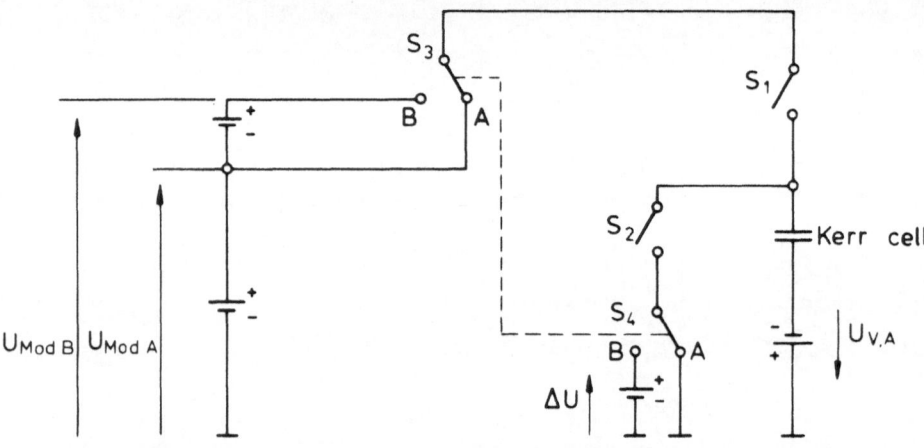

Fig. 12. Electronic circuit for the operation of a two-color deflector tube.

software and a few additional elements in the interface between the memory and the deflector electronics. This way, two-color laser projections have been extensively investigated. It turned out that even for rather closely spaced wavelengths as, e.g., yellow and green, symbols of different colors were clearly distinguishable.

DISPERSION COMPENSATION AND OTHER SYSTEM IMPROVEMENTS

For more than two laser wavelengths the use of separate deflection systems becomes somewhat cumbersome and it is desirable to pass at least two colors through one deflector. To do this, two dispersion effects have to be removed:

a)   The Kerr cell voltage varies noticeably with the wavelength of the modulated height, i.e., by about 15% for a change in wavelength from red to blue.

b)   The refractive indices of both calcite and nitrobenzene vary with wavelength. For nitrobenzene the variation is 1.3 % for the above wavelength difference and 0.7 % for the indices of calcite. Further, of importance is that also the difference of the ordinary and extra-ordinary index of calcite vary by approximately 4%. The first dispersion effect may be compensated in a rather straightforward fashion if light of one wavelength only is passed through the deflector at a time by connecting the high voltage switches S_1, S_2 of Fig. 3 to a second DC power supply. This may be done by switches essentially similar to the switches S_1, S_2. Fig. 12 gives an example of such a circuit.

The second dispersion effect is apparent in two ways. First, the absolute dispersion of the refractive indices of nitrobenzene and calcite will cause a drift of the total pattern of deflection directions. It is an easy matter to compensate for this effect by inverting the last birefringent prism for each deflection coordinate. This will, at the same time also compensate very effectively for drifts of the refractive indices with temperature (9). Whereas the operating temperatures for an uncompensated system would have to be kept constant to $\pm 0.03°C$, for the stability of a 1024 x 1024 position pattern to be better than 0.02 % the temperature may after introduction of compensation vary by as much as $\pm 4°C$. Second, this procedure does not compensate for the change of the difference of the refractive indices of calcite which determine the angular aperture of the total deflection pattern. Here, two types of compensations are possible: In a first method dispersion compensation is introduced through software design by correcting the position address. For position corrections of fractional line width, deflection stages of a correspondingly small splitting angle have to be added to the   deflector. In a second compensation method, the focussing objective O_1 of Fig. 13 may be designed so as to exhibit a dispersion of opposite sign. Computer calculations have shown that this can be done without introducing additional technical problems. The accuracy of superimposing the two position rasters may easily be brought to a fraction of a line width.

The absolute accuracy of the raster is sufficient for most display applications although it is not free of aberrations. These are introduced by the nonlinearity of Snell's Law of refraction. For small refraction prism angles and deflection directions subtending small angles at the system axis the deflection raster is of

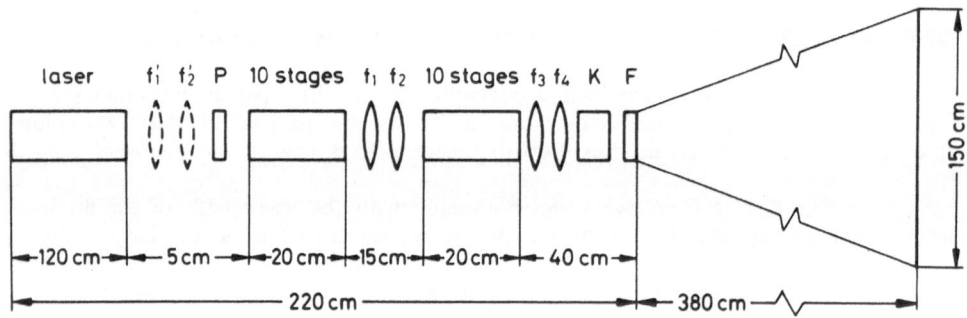

Fig. 13. Optical layout of a 20-stage deflector projection system.

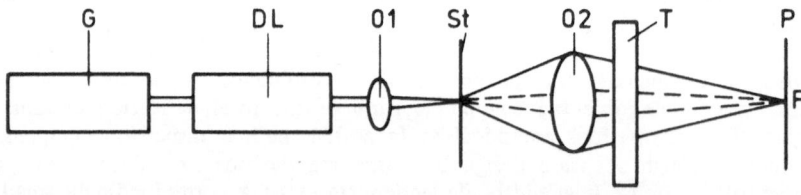

Fig. 14. Proposed system for removal of granulation.

excellent linearity. However, for a 1024 x 1024 position raster generated by a deflector the largest deviation of a position from the theoretical value - at the rim of the raster - corresponds to about 1 elementary position distances. Further improvements are available by composing each refraction element of two or more single birefringent prisms and by orienting the prisms individually in the plane of refraction.

Another problem especially associated with laser displays is that of granulation which is caused by the high degree of coherence of laser light. Each laser-illuminated element of the screen appears twinkling to the observer upon the slightest motion of his eyes. Many methods have been suggested to remove this well known and often described phenomenon (10, 11). The general approach is to generate a large number of statistically independent granulation patterns at such a high rate that the eye notices an averaged intensity distribution in effect equal identically to that from thermal light sources. Theory shows that 100 different granulation patterns have to be superimposed for an intensity oscillation of the "twinkling effect" of not more than 10 %. The solution to this problem is especially difficult for scanned displays since all 100 independent granulation patterns have to be generated during the dwell time of the laser beam, i.e., within a few microseconds. Fig. 14 gives a schematic diagram of a device meeting this requirement and suitable for multicolor displays. The zero mode laser beam is focussed in the primary image plane P_1 onto a scattering screen, which converts the single mode into a large number of spatial modes being focussed by the objective lens O_2 onto the display screen at a point corresponding to the primary image point. A plane parallel refractive transparent medium that is being excited by an ultrasonic wave modulates the phases of the various modes essentially in random fashion. The degree of averaging of the granulation depends on the number of spatial modes of frequency and amplitude and of the piezooptic properties of the medium. Experiments have shown good results.

CONCLUSIONS

The experiments described above have shown that digital laser displays are well suited for the presentation of alphanumeric and graphic information using commercially available laser sources. The digital character of the electronic drive circuitry makes the control by any digital data system a simple affair. Multicolor, real time and on-line manipulations of large screen displays of high brightness are some of the prominent features of this technique which still leaves space for many improvement especially regarding resolution and scanning speed. 22-stage deflectors are presently being tested (12). Traffic control could become one of the major application areas of digital laser displays.

REFERENCES

(1)   U. J. Schmidt: The problem of light beam deflection at high frequencies, in optical processing of information, D. K. Pollock, C. J. Koester and J. T. Tippett, eds., Baltimore, Spartan Books, 1963, pp. 101-103.

(2)   T. J. Nelson: Digital light deflection, Bell. Syst. Tech. J., vol. 43, pp. 821-845, March 1964.

(3)   W. Kulcke, K. Kosanke, E. Max, M. A. Habegger, T. J. Harris and H. Fleisher: Digital light deflectors, Appl. Optics, vol. 5, pp. 1657-1667, Oct. 1966.

(4)   H. Meyer, D. Riekmann, K. P. Schmidt, U. J. Schmidt, M. Rahlff, E. Schröder, and W. Thust: Design and performance of a 20-stage digital light beam deflector, Appl. Optics, vol. 11, pp. 1732-1736, Aug. 1972.

(5)   U. J. Schmidt, E. Schröder, and W. Thust: Optimization Procedures for digital light beam deflectors, Appl. Optics, vol. 12, pp. 460-466, March 1973.

(6)   U. Krüger, R. Pepperl, and U. J. Schmidt: Electro-optic materials for digital light beam deflector, Proc. IEEE, vol. 61, No. 7, pp. 992-1007.

(7)   U. J. Schmidt: Electro-optic light beam deflection, to be published in Philips Technical Review.

(8)   W. Thust: A computer-controlled graphic laser display, 1972 SID International Symposium, San Francisco, June 6 - June 8, 1972, Digest of technical papers, pp. 156-157.

(9)   U. J. Schmidt and W. Thust: Temperature stabilization of the deflection pattern of a digital deflector containing single prisms, J. Opto-Electronics, vol. 2, pp. 29-35, 1970.

(10)   H. Arsenault and S. Lowenthal, Opt. Commun., vol. 1, P. 505, 1970.

(11)   E. Schröder: Elimination of granulation in laser beam projection by means of moving diffusers, Opt. Commun., vol. 3, pp. 68-72, 1971.

(12)   E. Schröder: Realisierung eines elektro-optischen Lichtablenkers mit einer Auflösung von mehr als 10^6, Proc. Seminar Elektro-Optik, Munich, Jan. 1972.

TRANSPARENT CONDUCTIVE COATINGS

E. Ritter

Balzers AG

Balzers (Fürstentum Liechtenstein)

1.   INTRODUCTION

Thin films are useful components in electro-optical devices.

Examples are laser mirrors, monochromatic interference filters, neutral and dichroic beam splitters, antireflection coatings and transparent conductive coatings (TCC).

It is the purpose of this contribution to list suitable materials and deposition techniques for transparent conductive coatings and to describe the properties and some of the applications of these films.

2.   MATERIALS

TCC combine high optical transmission with good electrical conductivity. The existence of both properties in the same material is from the physical point of view not trivial and is only possible with certain semiconductors like indium oxide, tin oxide, cadmium oxide and, to some extent, with thin gold films.

The pure oxides of Cd, In and Sn should exhibit insulating properties. But due to the existence of nonstoichiometric composition they show in many cases semiconducting behaviour, oxygen vacancies causing the conduction.

The reproducibility of the creation of oxygen vacancies and the stability of these vacancies, especially at higher temperatures, is not very good. It is therefore of advantage to create donor or acceptor states by doping the material.

The dopant normally used for In_2O_3 is tin. The tin atoms replace In

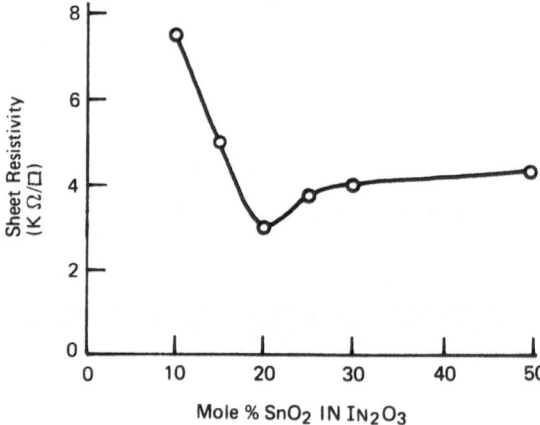

Fig. 1. Sheet resistance versus target composition for 1000 Å thick films of In_2O_3/SnO_2 deposited in pure oxygen (Vossen, ref. 1).

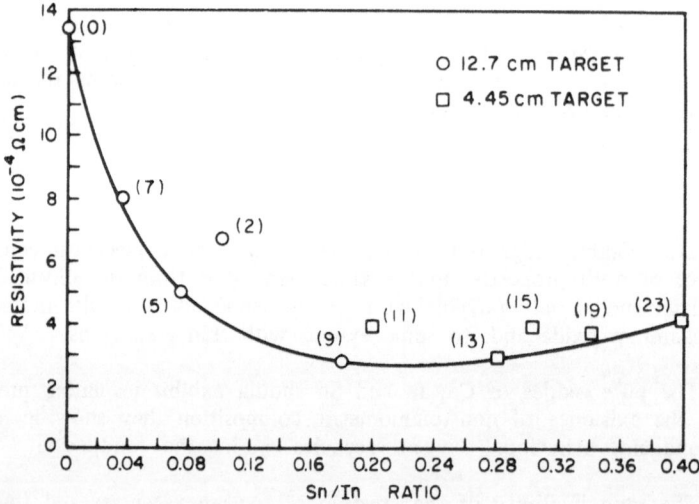

Fig. 2. Resistivity of In_2O_3 films containing varying amounts of SnO_2 plotted as a function of the Sn/In ratio measured by x-ray fluorescence. The numbers in parentheses indicate nominal SnO_2 mol% content of the targets used (Fraser and Cook, ref. 2).

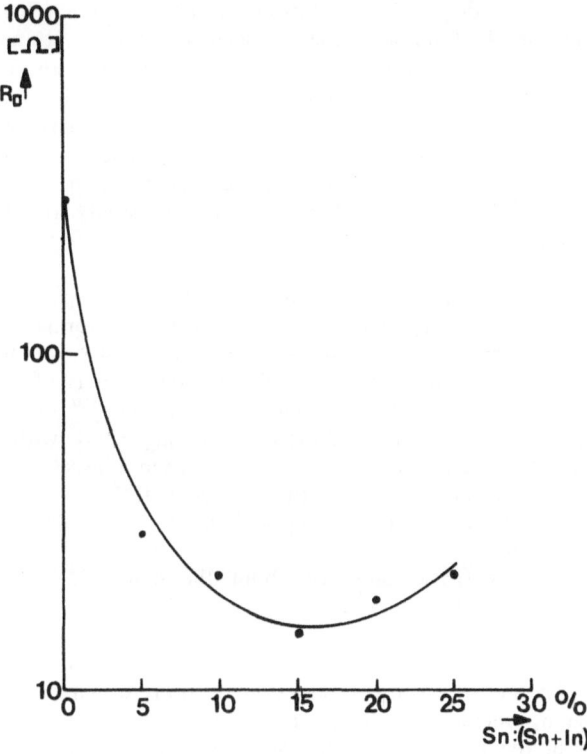

Fig. 3. Sheet resistance as a function of the ratio Sn/In in the target (Sager, ref. 3).

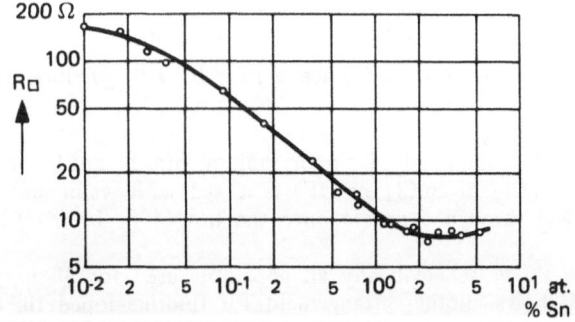

Fig. 4. Sheet resistance of indium oxide films in dependence on tin doping. Film thickness 3200 Å (von Boort and Groth, ref. 4).

atoms in the lattice and deliver free electrons to the conduction band due to their higher valency. The level of doping varies somewhat but the optimum seems to lie around 80 mol% In and 20 mol% Sn for sputtered films. Figures 1 - 4 show examples for films deposited by sputtering of the oxides and by reactive sputtering of an In-Sn alloy and by chemical vapour deposition. In the last case the doping level is lower, at least in the starting material (2-3 %). The tin content in the film may be higher due to different condensation coefficients. With this doping level carrier concentrations of about $5 \cdot 10^{20}$ cm^{-3} and a resistivity of about $2 \text{-} 10 \cdot 10^{-4}$ Ω-cm are achievable.

In the case of tin oxide antimony is used as a dopant in the form of Sb_2O_3 or $SbCl_3$ in the starting material. This leads to the formation of a controlled valence semiconductor, each Sb atom in its valence state Sb^V donating one electron to the conduction band of the lattice (5). The doping level is of the order of 0.2 -1.0% (6) and causes a carrier concentration of about $5 \cdot 10^{20}$ cm^{-3} (4). The same carrier concentration can also be achieved by doping SnO_2 with fluorine (7). Fluoride ions replace oxygen ions and since they need one electron less than oxygen to fill the 2p orbital for every fluoride substitution a tin atom retains an extra 5s electron which enters the conduction band of the lattice.

Also here resistivity values of about the same order as in the case of tin-doped indium oxide can be achieved.

From the optical point of view indium and tin oxides are preferable since their band gap ~3.5 eV allows the production of films which are (fully) transparent in the visible region, whereas CdO has a band gap of about 2.5 eV, leading to appreciable absorption in the blue part of the visible spectrum (Fig. 5) (8).

The transmission beyond the absorption edge in the region of high transmission depends on the degree of doping and decreases with increasing conductivity. This will be discussed later on.

3.   METHODS OF DEPOSITION

3.1   Chemical Vapor Deposition-Spraying Technique

This technique was used since a long time to produce TCC of SnO_2 (9).

a)   Principle
A volatile compound of tin or indium, mostly $SnCl_4$ or $InCl_3$, together with the dopant ($SbCl_3$ or $SnCl_4$ or HF) is mixed with water and often an organic solvent and sprayed through a nozzle onto the hot (400 - 1000°C) substrate.

Fig. 6 shows schematically an apparatus used for Sb-doped tin oxide films (6), Fig. 7 shows another arrangement for fluorine-doped tin oxide films (7).

It was found to be necessary to control the temperature of the reactive gas mixture as well as of the substrate as closely as possible to obtain reproducible results (7).

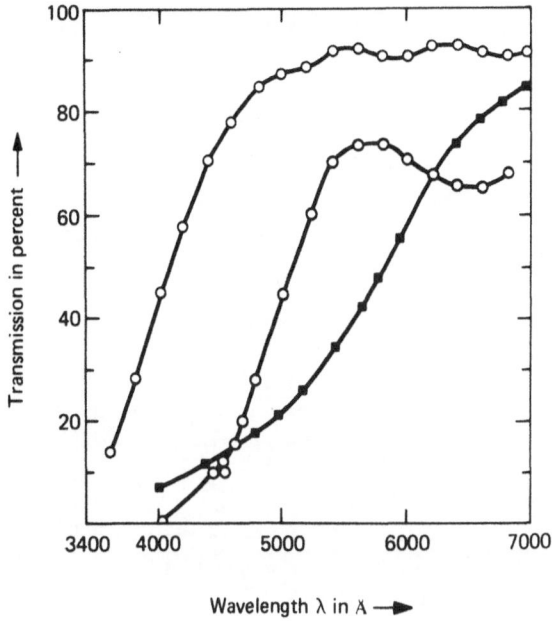

Fig. 5. Spectral transmittance of CdO and In_2O_3/SnO_2 films. Curve 1 is for CdO sputtered in 100% O_2 ambient; curve 2 is for CdO sputtered in 2.5% O_2 in N_2 ambient; curve 3 is for In_2O_3/SnO_2 (Mehta and Vogel, ref. 8).

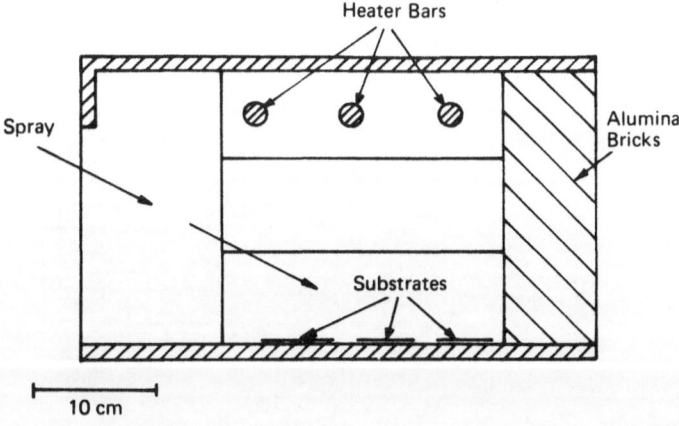

Fig. 6. Section through a spraying oven (Peaker and Horsley, ref. 6).

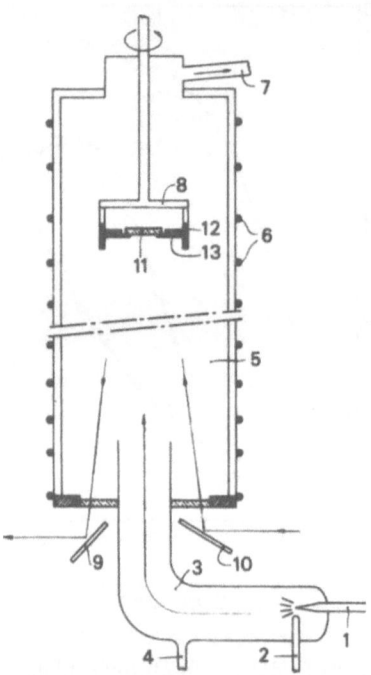

Fig. 7. Apparatus for the production of fluorine doped tin oxide films (1 and 2 nozzles for the reactive gas, 4 return of non used liquid, 5 chamber, 6 heating elements, 7 gas outlet, 8 substrate holder, 9 and 10 mirrors for visual observations, 11 substrate, 12 stainless steel support, 13 platinum support (Groth, ref. 7).

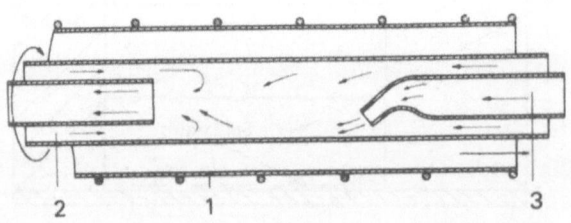

Fig. 8. Arrangement for the coating of inside walls of tubes. 1 Heating elements, 2 tube, to be coated, 3 nozzle for the reacting gas (Groth, ref. 7).

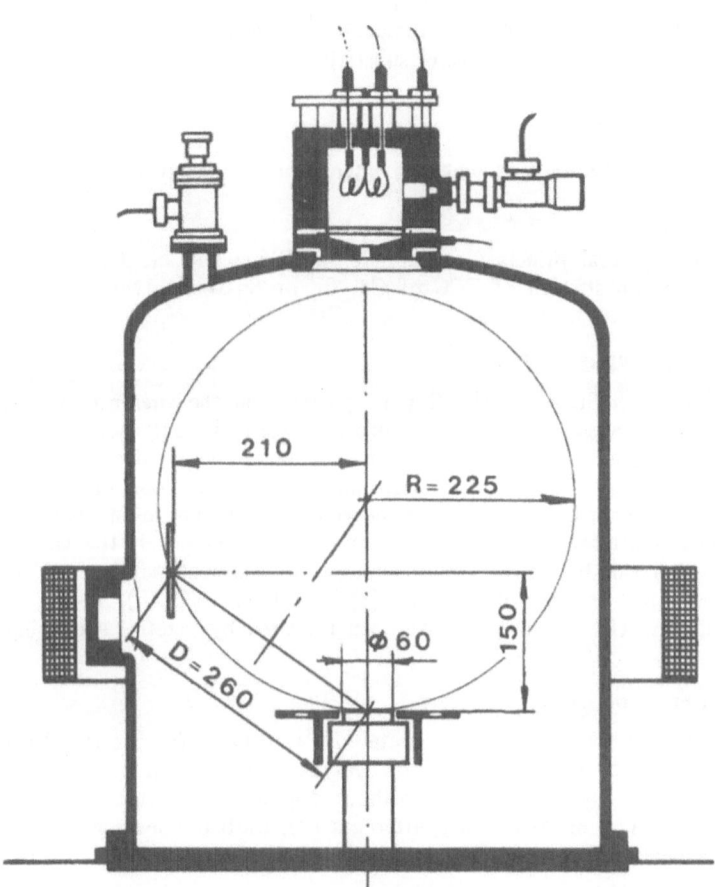

Fig. 9. Sputtron arrangement (Sager, ref. 3).

b)   Discussion
          Positive aspects are the good film quality with respect to adherence and electrical and optical properties, and also the relatively inexpensive equipment and the independence of the shape of the substrate. Even inside walls of tubes can be coated (7) (Fig. 8).

          Negative factors are problems of thickness control and thickness homogeneity and the very high substrate temperature, which excludes high quality glass with high surface flatness and plastic substrates.

3.2.   Sputtering

          This is the method which is now most frequently described in literature.

a)   Principle
          The general principle of cathode sputtering will be presupposed here. For the case of production of TCC of Cd, In, or Sn oxide two special methods may be used:

α) Reactive sputtering of metal

Cd, In, Sn or an In-Sn or Sn-Sb alloy is sputtered in the presence of oxygen to achieve oxidation during the sputtering process. Normally only some oxygen is added to the sputtering gas to avoid oxidation of the target which decreases the sputtering rate. The reactive sputtering can be performed with a dc diode or triode arrangement. As an example the arrangement for triode sputtering of an In-Sn alloy is shown in Fig. 9 (called Sputteron). The advantage of this arrangement is the large number of substrates which can be coated simultaneously. It was successfully used for the preparation of tin-doped indium oxide films (3). Films of CdO (8), In_2O_3 (SnO_2 doped) (8) and SnO_2 (Sb_2O_3 doped) (10) have been prepared by reactive sputtering by other authors.

β) HF-sputtering of oxide

Instead of reactive sputtering of the metals HF-sputtering (ev. normal DC-sputtering) of the oxides can be used. This method was used for indium oxide (tin oxide doped) (1, 2, 11) and for In_2O_3 and SnO_2 without doping. Best results were obtained with pure noble gases (Ar or Xe) as a sputter gas (2), probably leading to additional oxygen vacancies in the oxide lattice.

b)   Discussion
          Positive aspects of the sputter method are the good reproducibility of doping levels, the lower substrate temperature and the desirable film properties. The substrate temperature is often determined by an annealing process at $\sim 200^{\circ}C$ during or subsequent to the sputtering process which improves the transmission, conductivity and stability of the films (Fig. 10). Annealing at very high temperatures in air on the other hand may lead to a pronounced increase in resistance (3).

          Negative: For the homogeneous deposition on large substrates special installations are necessary. The equipment is relatively expensive. Also the target preparation poses some problems.

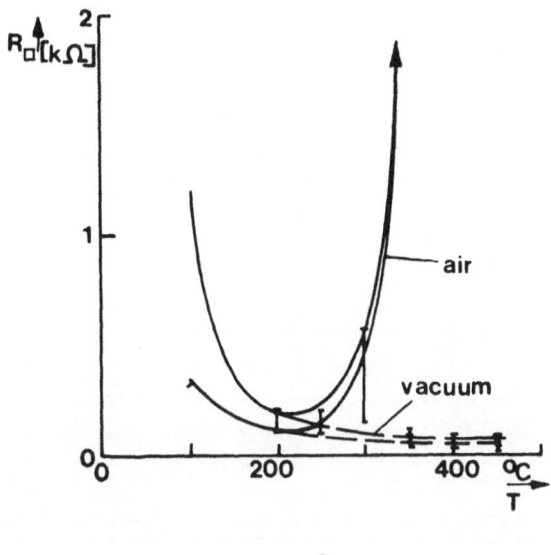

a

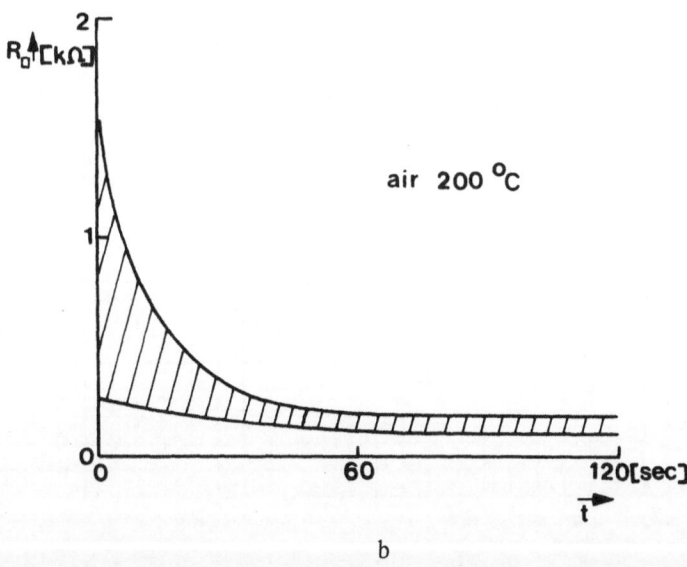

b

Fig. 10. a) Dependence of sheet resistance on annealing in air at 200°C and b) sheet resistance as a function of temperature after 30 minutes of annealing (Sager, ref. 3).

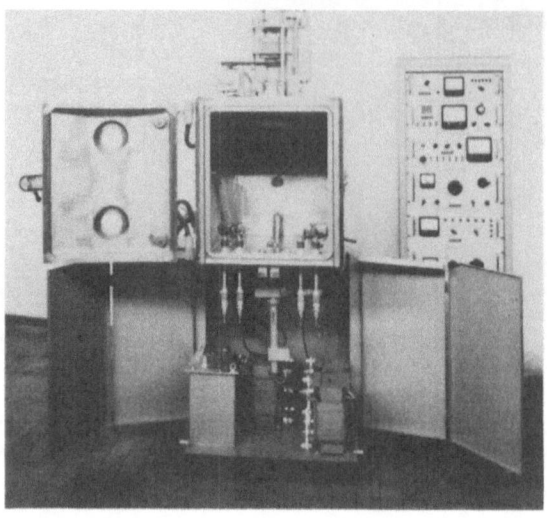

Fig. 11. Evaporation plant BA K 550 (Courtesy of Balzers AG).

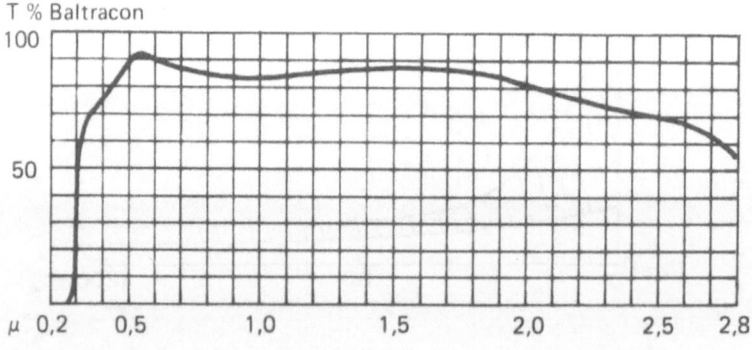

Fig. 12. Spectral transmission of Baltracon (tin-doped indium oxide film) (Courtesy of Balzers AG).

3.3.   Evaporation

The principle of vacuum evaporation is also presupposed here. Fig. 11 shows an evaporation plant suitable for the production of TCC.

A direct evaporation of the oxides in question is not possible since they all tend to dissociate.

Possible solutions:
a)   Evaporation of the oxides with subsequent oxidation by heating in air or by some other (chemical or electrochemical) methods (12).

b)   Reactive evaporation of the metal or alloy in an oxygen atmosphere (also in some cases followed by an oxidation step to complete the oxidation). Our films on glass substrates are not completely oxidized in order to facilitate the etching. In this state they can be etched with hydrochloric acid only, whereas in the fully oxidized state a Zn dust catalyst is necessary.

The resistance of the freshly evaporated films is higher than that of the annealed ones. Also they show ~ 30 % absorption. The annealing is performed at $350^{\circ}C$ for 30 minutes.

Discussion:
a)   Positive: Very low substrate temperature possible ($< 100^{\circ}C$), therefore coating of plastics possible. Large areas can be coated with good thickness control and thickness uniformity.

b)   Negative: Expensive equipment necessary. Control of doping level more difficult than with sputtering.

4.   PROPERTIES OF THE FILMS

4.1.   Electrical Properties
It is common practice to specify resistance in this case in Ω_o, which uses the fact, that the resistance of squares is independent of absolute size. The advantage of using area resistance is that no thickness has to be specified.

For CdO films Ω_o values of 26 - 82 are reported (8). For SnO_2 films (F-doped) the lowest values are around 15 Ω_o (7), for In_2O_3 films (Sn-doped) 1 - 10 Ω_o (2). Commercially available In oxide films range between 10 and 1000 Ω_o, SnO_2 films between 30 and 1500 Ω_o.

4.2.   Optical Properties
Films with an area resistance of 50 - several hundred of indium oxide (tin doped) show only an absorption of 1-2 % in the visible range (Fig. 12). The fluctuation in transmission observed are due to an interference effect and not to absorption. That means low transmission is caused by high reflectance (Fig. 13). Figure 14 shows as an example the transmission of 5 films with different optical thickness (3) ($\lambda/4$, $\lambda/2$, 3 $\lambda/4$, λ, 5 $\lambda/4$; figure of merit $\sim$ 1). With higher doping levels and higher

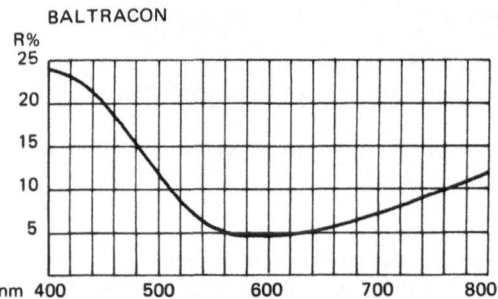

Fig. 13. Spectral reflectance of Baltracon (Courtesy of Balzers AG).

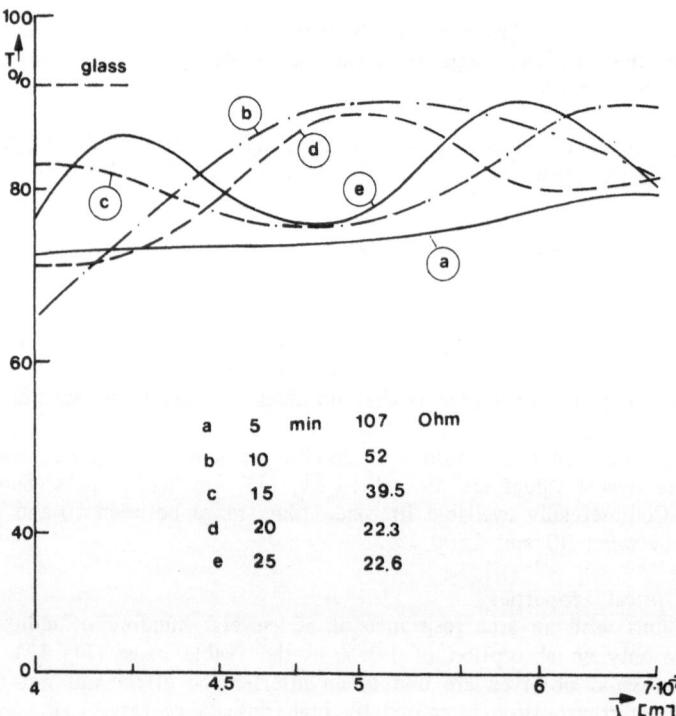

Fig. 14. Spectral transmission for films of different thickness (Sager, ref. 3).

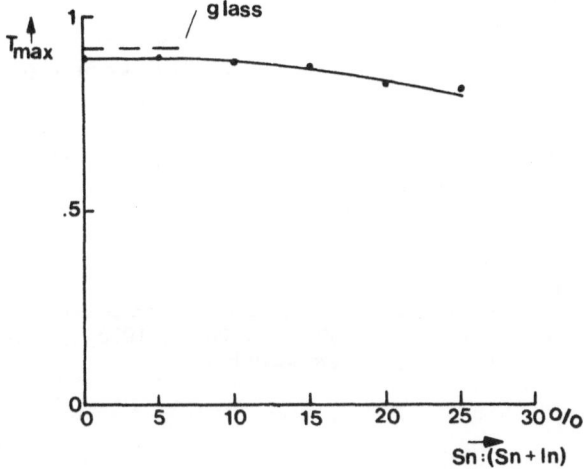

Fig. 15. Peak transmission of a $3\lambda/4$-film as a function of doping level (Sager, ref. 3).

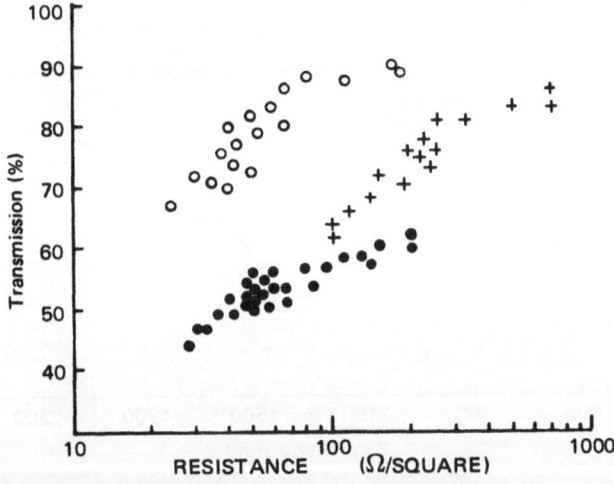

Fig. 16. Variation of optical transmission with film resistance for various doping conditions: + no antimony, o 0.4 mol % antimony, ● 2 mole % antimony (Peaker and Horsley, ref. 6).

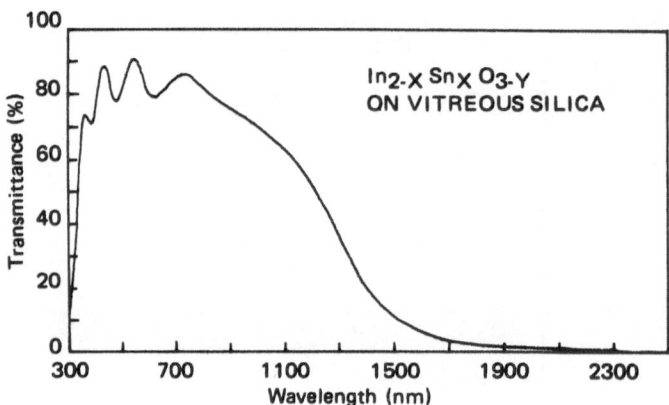

Fig. 17. Spectral transmission of an $In_{2-x} Sn_x O_{3-y}$ film sputtered from an $In_2O_3 - 9$ mol %
SnO_2 target (Fraser and Cook, ref. 2).

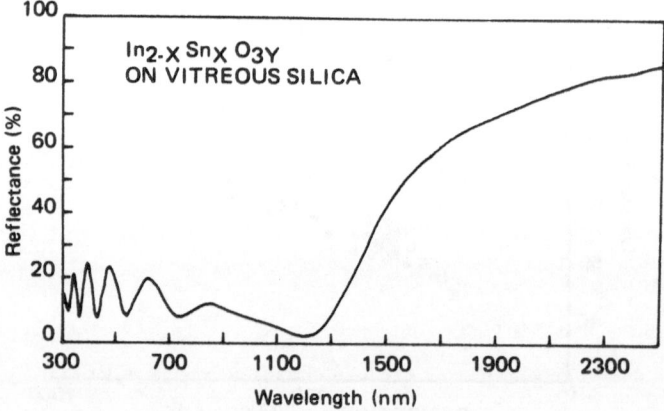

Fig. 18. Spectral reflectance of an $In_{2-x} Sn_x O_{3-y}$ film sputtered from an $In_2O_3 - 9$ mol %
SnO_2 target (Fraser and Cook, ref. 2).

conductivity the transmission drops somewhat but the figure of merit (% T/R) still increases up to values as high as 20 (Fig. 15, 16, 17). Around 1μ a strong absorption appears.

With increasing conductivity the reflectance in the infrared also increases strongly, reaching values of 80 % at 4 μ, which makes such films very useful as heat reflectors for sodium vapor lamps (Fig. 18, 19).

About the same facts are valid for tin oxide films (Sb or F doped) (Fig. 20).

4.3.   Mechanical and Chemical Properties
Tin oxide films are very hard and scratch resistant. They can be easily cleaned with all common cleaning fluids. Chemical etching is only possible with $HCl + Zn$ dust catalyst.

Indium oxide films are somewhat less hard, depending on the SnO_2-content, but still quite robust. Also these films can be cleaned easily. Pure indium oxide films or films with low tin doping level can be etched with HCl only, as well as not completely oxidized In_2O_3-SnO_2 films. Heavily doped films which are completely oxidized may require the Zn dust catalyst for etching.

4.4.   Temperature Stability
Doped tin oxide coatings and doped indium oxide coatings, especially prepared by chemical vapour deposition or vacuum evaporation, show very good temperature stability. Prolonged heating at $500°C$ does not change the resistance value appreciably. Undoped films as well as some sputtered films show a pronounced increase of resistance at temperatures above $300°C$. (Fig. 10). This is probably due to the disappearance of oxygen vacancies and to the lack of a protective SnO_2-rich cover layer.

4.5.   Structure
Tin oxide films are tetragonal (10), indium oxide films cubic (bcc) (1,2). Often a pronounced texture is observed (2) (Fig. 21).

5.   APPLICATIONS.

5.1.   Liquid Crystal Displays

For small displays (20 × 30 mm) as for instance wrist watch displays, normal drawn sheet glass with a flatness of ~ 5 μ/cm is usable. For larger displays the requirements are much more stringent, typical values being 2 μ/cm. This flatness figure can be met by float glass, but only with a thickness of 3 mm or more. For some applications even a flatness figure of 0.2 μ/cm is asked for, which can only be achieved by optical polishing.

For purely reflective displays aluminum mirrors can be used as a back electrode or the aluminum can be deposited on the rear side of microsheets. For displays which work in reflection as well as in transmission a dielectric mirror is wanted.

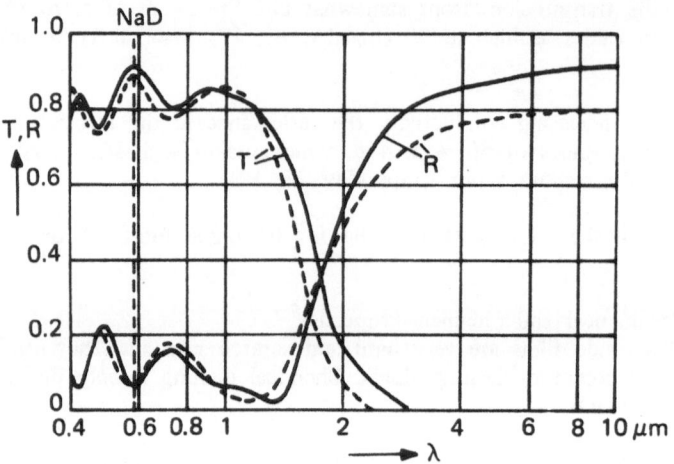

Fig. 19. Solid line: Spectral transmission and reflectance of a tin-doped indium oxide film. Dashed line: Spectral transmission and reflectance of an antimony-doped tin oxide film (von Boort and Groth, ref. 4).

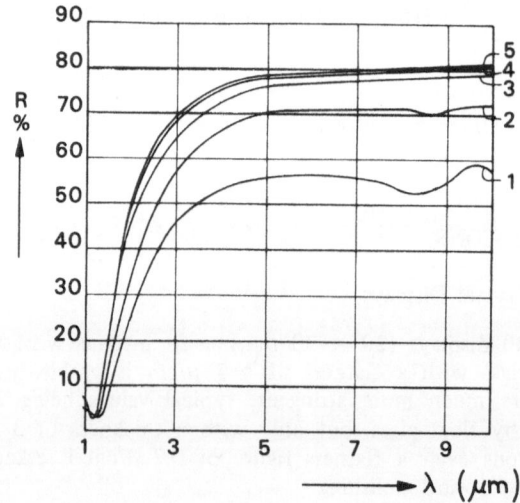

Fig. 20. Spectral reflectance of a fluorine-doped tin oxide film (Groth, ref. 7).

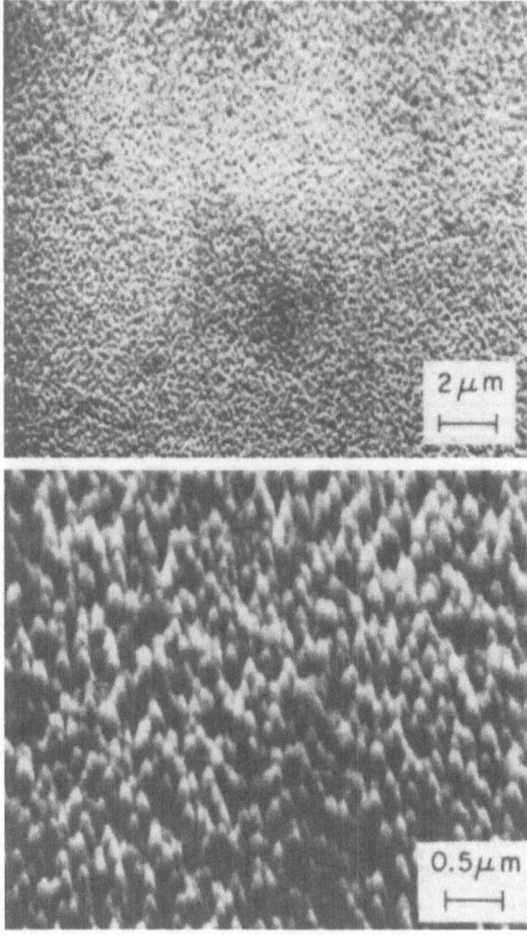

Fig. 21. Scanning electron micrographs of an $In_{2-x} Sn_x O_{3-y}$ film sputtered with Ar only from an $In_2O_3 - 9$ mol% SnO_2 target (Fraser and Cook, ref. 2).

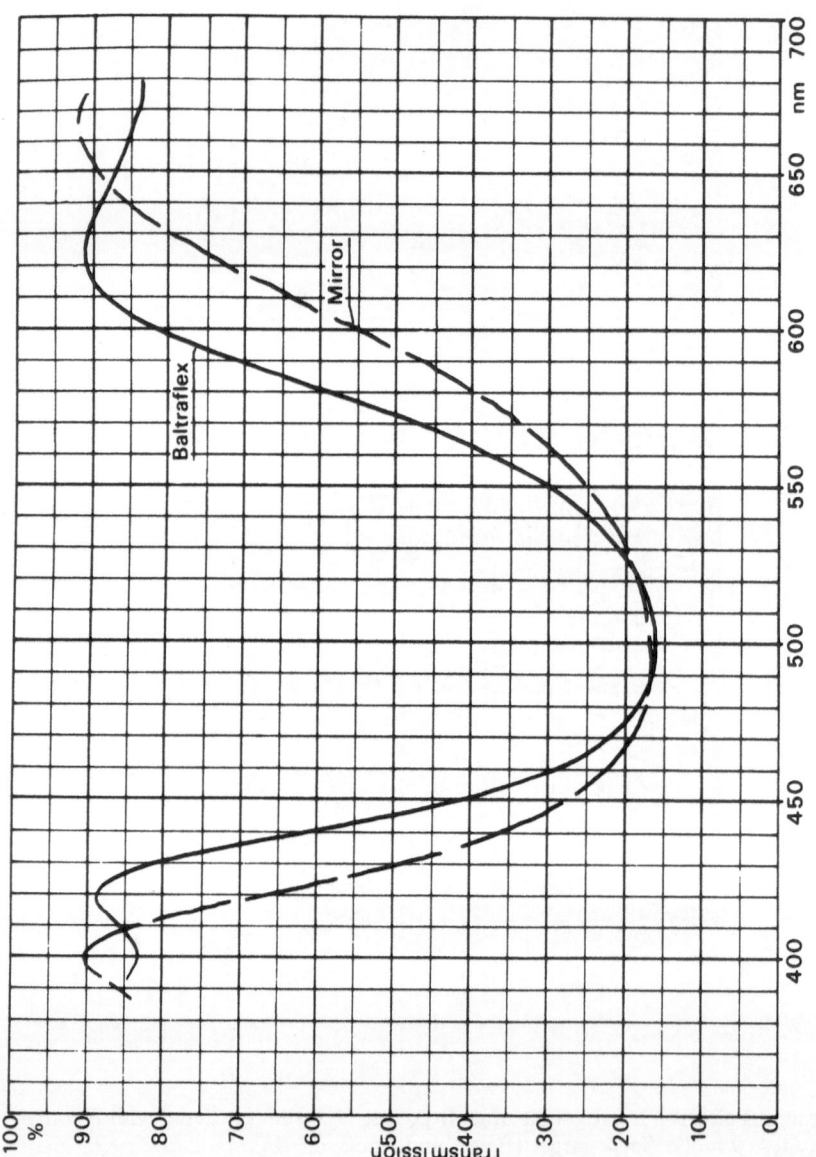

Fig. 22. Spectral reflectance of a combination of dielectric mirror plus transparent conductive coating (Baltraflex) and of the dielectric mirror alone (Courtesy of Balzers AG).

We produce such combinations of dielectric multilayer stacks + TCC under the trademark BALTRAFLEX. They consist of a dielectric stack of 5 - 7 layers + TCC. The reflectance value and the wavelength position of the maximum reflectance can be adjusted at will. Therefore coatings with blue, green or yellow-red reflection colors can be produced. Fig. 22 shows an example.

In order to reduce the visual contrast between etched and un-etched areas of the coating we tried to make the difference in optical characteristics. This can be seen in Fig. 22, where the transmittance of BALTRAFLEX is compared with the transmittance of the mirror coating alone, where the TCC has been etched away. The difference is relatively small.

In order to improve the orientation of liquid crystal materials, other methods besides evaporation seem to help. SiO, MgF_2 or a metal like copper is evaporated under very oblique incidence on the substrate with the finished pattern. Only a very thin film (100 Å) is required. Such a coating improves the orientation of liquid crystal materials, but is quite sensitive to cleaning and touching and should therefore only be applied immediately before cell mounting. Possibilities of improving the stability of this coating are now under study in our laboratory.

5.2.   Electroluminescent Displays

Low resistance values required ($< 30\ \Omega_o$). Flatness not so critical.

5.3.   Gas Discharge Displays

Good temperature stability is a must! Flatness not so critical.

5.4.   Heating Elements (Windows for Aircraft, etc.)

Very low resistance values necessary ($< 10\ \Omega_o$). Often thin gold coatings! (70 - 80 % T, 1 - 10 Ω_o)

5.5.   Antistatic Coatings for Meters

Uncritical. Can be met by undoped indium oxide.

5.6.   Pockels Cells for Laser Q-Switch

REFERENCES

(1)   J. L. Vossen, RCA Review, 32, 289 (1971).

(2)   D. B. Fraser and H. D. Cook, J. Electrochem. Soc. 119, 1368 (1972).

(3)   O. Sager, in "Deposition Technique for Contacts and Inter-connections

on Integrated Circuits" (edited by Balzers AG) Balzers AG, Balzers 1972.

(4)   H. J. J. von Boort and R. Groth, Phil. Techn. Rundschau, 29, 47 (1967).

(5)   C. A. Vincent, J. Electrochem. Soc., 119, 515 (1972).

(6)   A. R. Peaker and B. Horsley, Rev. Sci. Instr. 42, 1825 (1971).

(7)   R. Groth, N. V. Philips Gloeilampenfabrieken, Auslegeschrift 1496
        590 (Oct. 16, 1964).

(8)   R. R. Mehta and S. F. Vogel, J. Electrochem. Soc., 119, 752 (1972).

(9)   Libbey-Owens-Ford Glass Co., and H. A. Mac Master, Brit. Pat. 632 256
        (1942).

(10)   E. Leja, Acta Physica Polonica, 1738, 165 (1970).

(11)   J. R. Boswell and R. Waghorne, Thin Solid Films, 15, 1415 (1973).

(12)   Teijin Ltd. and S. Sobajama et al., Offenlegungsschrift 2, 255, 561,
        (Nov. 13, 1972).

DISPLAY DRIVE CIRCUITS AND THEIR IMPACT ON SYSTEM ECONOMICS

P. B. Page

ITT Components Group Europe Central Applications Laboratory Harlow

Harlow, Essex (U. K.)

1. INTRODUCTION AND BACKGROUND

Once the initial excitement of a new component research breakthrough has subsided and the new devices start to become available in quantity the system designer is faced with the familiar task of producing a system to meet a specification with tight economic constraints. This is the situation that exists now in the display field with light-emitting diodes (LED) and is beginning to develop with liquid crystal cells (LCC).

With any fundamentally new device two basic questions can be asked:

1.1. Is its performance/cost ratio better than that of the existing devices?
The weightings given to performance and cost, of course, depend on the nature of the application.

1.2. Can it perform a function that has not been practical before?
Often it takes considerable time for the potential of a new device to be seen. The ability of device designers and systems engineers to communicate fully is a key factor in the rate of development of new applications.

With LED and LCC displays the first question is answered in part by comparing them with existing devices such as the cold-cathode gas tube, the phosphor anode tube and the incandescent tubes. Wrong conclusions can be drawn, however, if the impact of the cost of the associated circuits are not also taken into account.

The basic functions to be performed by the circuits can be summarized as follows:

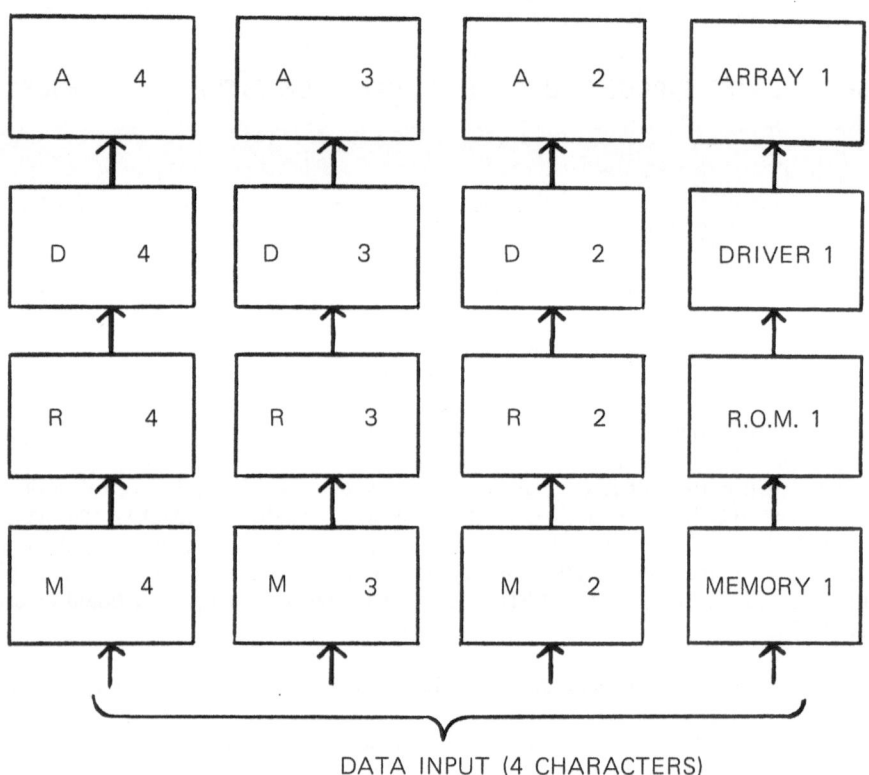

Fig. 1. Basic static display system.

Character Selection - to route the input data to the required output device.

Character Generation - to convert the binary coded input data to the form essential to drive the display.

Data Memory - to store the input data, freeing the input system to perform other functions.

Clock Control - to synchronize the various parts of the system.

Display Drive - to provide signals at the right voltage and current levels to operate the display.

The trend in system design to the use of integrated circuits (IC) especially large scale integrated circuits (LSI) sets certain constraints on the techniques that can be used to implement these functions. The major constraints are:

(i)   limited drive voltages and currents - cost rises and reliability falls as these are increased.

(ii)   the need for minimum interconnections between the display module and the system-package-costs rise and yields fall as the number of pins are increased.

(iii)   the need for small module size for a given character size integration means that systems like pocket calculators can now be very small; thus the display must be compatible.

Traditionally the above circuit functions have been implemented in two basic ways:

A.   Static System: in which all the functions are duplicated for each character in the display. This has the benefits of modularity but can be uneconomic for some applications. Alphanumeric displays are a case in point where a high percentage of the circuit cost is in the character generation and the display drive (see Figure 1).

B.   Time-Shared System: in this case the character generation and display drive functions are shared amongst a group of characters by dynamic multiplexing techniques. Various techniques exist for implementing this system but all suffer from a lack of modularity and, because the displays are pulsed, suffer brightness and peak current problems with LED's (light emitting diodes) and operating speed limitations with LCC's (liquid crystal cells) (see Figure 2).

Time-Shared Displays: With time-sharing, only part of the display is on at any one time. To avoid significant flicker the whole display must, however, be refreshed at least every 20 mS. Also, to achieve the same brightness as a static system the current pulse in a LED array must be greater than the normal d.c. value. In general the amplitude of the current pulse will be nI_{dc}, where n is the number of scan points and I_{dc} is the normal d.c. current. Figure 2 shows the three basic forms of the time-sharing systems as applied to a 5 x 7 dot matrix display device.

a)   With the first, the column data from the ROM (Read-Only-Memory) Character Generator IC which typically stores the 5 x 7 matrix patterns of 64

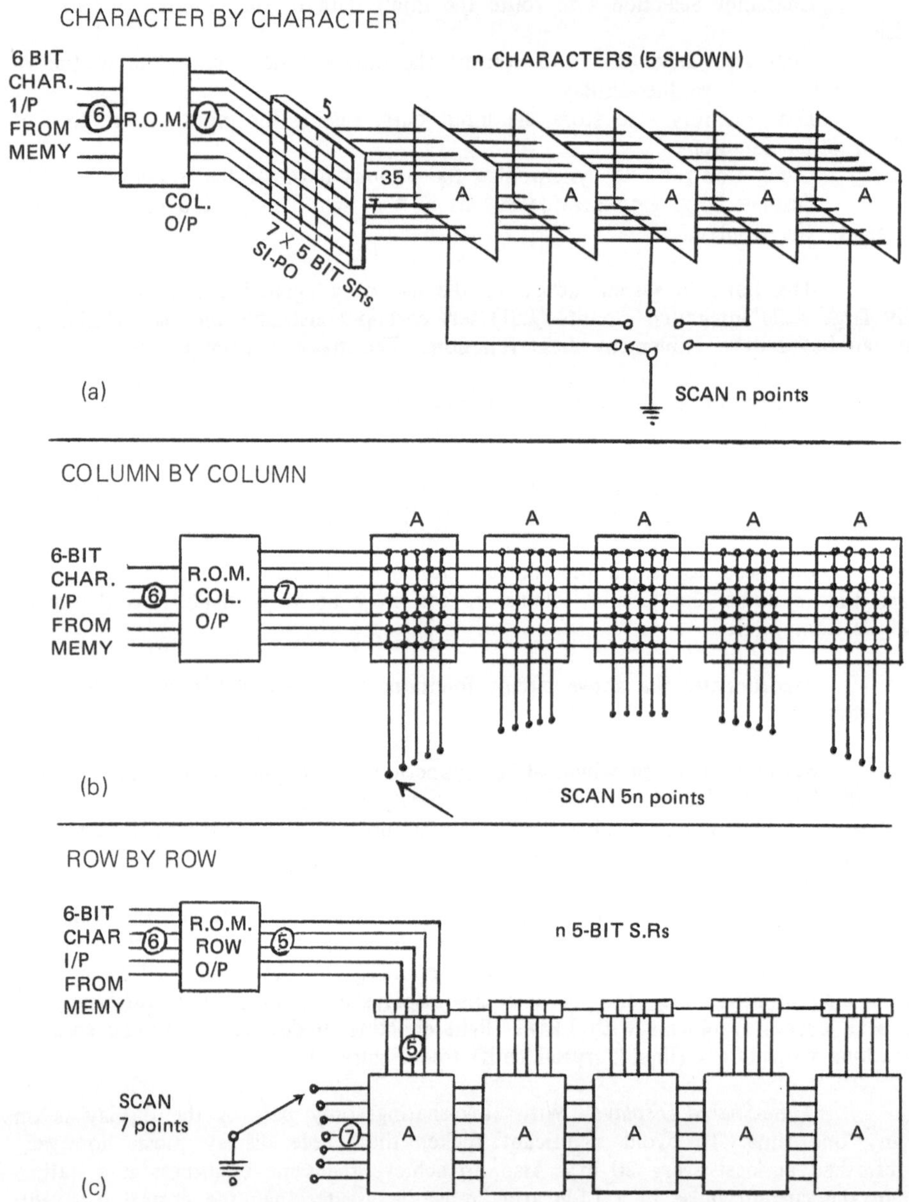

CHARACTER BY CHARACTER

(a)

COLUMN BY COLUMN

(b)

ROW BY ROW

(c)

Fig. 2. a) Time-shared display character by character; b) time-shared display column by column; c) time-shared display row by row.

characters) is built up in seven 5-bit serial-in/parallel-out shift registers. The 35 out-puts from the shift registers are then applied, via drivers, simultaneously to the 35 positive connections on the arrays, and the common negative connection of one array only is connected to ground via a transistor in the scanner. The binary address to the ROM is then changed and the process repeated except that the scanner now connects the next array in the sequence, to ground. This is the only technique that can be used with monolithic LED arrays as one side of all the diodes in a character are commoned by the semiconductor substrate.

b)   The second technique requires an array with an X-Y address rather than a common negative plus 35 as in "a". All like rows are connected together thus the ROM column outputs via seven drivers can feed display without a "build-up" memory. Scanning is more complicated, however, as five times as many scan steps are required. Also the current pulse is five times greater as only a column of each character is on at any one time.

c)   The third technique also requires an X-Y addressed array but scans on the rows and applies data to the columns. Only seven points are scanned for any length of array, which is a great advantage, but 5 bits of build-up memory are required per character. Data from a row output ROM is parallel loaded then serially shifted until the data from the top row of all characters is stored. The scanner then connects the other side of the top row elements and display is achieved, the process being repeated for the other six rows.

        In general method "a" is preferred for up to six-characters and method "c" for seven or more. Method "b" is not of interest with LED or LCC display devices.

2.   COMPATIBLE INTEGRATED DRIVE CIRCUITS FOR LCC AND LED
        DISPLAYS

        To exploit fully the benefits of LED and LCC devices and to meet the constraints set out above the preferred solution in general terms is to incorporate a special IC on the substrate with the display and perform the interconnection and packaging using normal IC techniques. This approach, however, cannot be fully ap-plied with either of the systems A or B.

        With the "Static System" the circuit can be reduced to one IC but the IC is specific to the type of display font i.e. a different IC is required for 7-bar numeric and 5 x 7 dot matrix alphanumeric. Thus the need to develop and stock a range of IC's is added to the other problems listed above.

        With the "Time-Shared System" the lack of modularity of the circuits and the need (with LED displays) to switch high currents makes integration very complex.

        It is essential to make the circuit concepts for both LED and LCC displays as similar as possible as this will help reduce cost by increasing the pro-duction quantities of the particular IC.

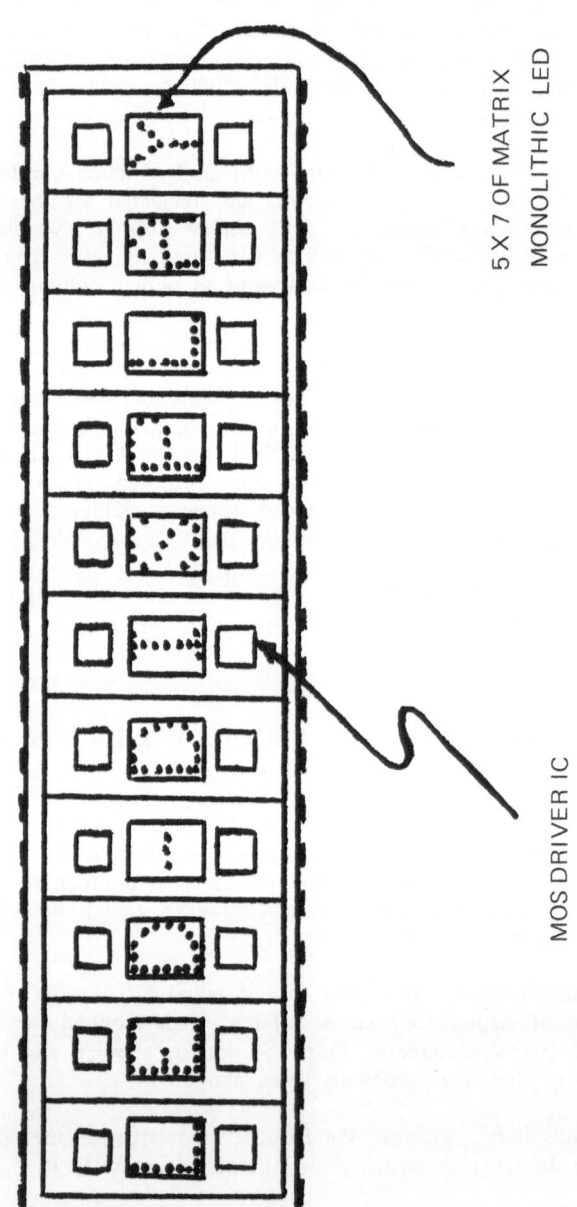

5 X 7 OF MATRIX
MONOLITHIC LED

MOS DRIVER IC

Fig. 3. Alphanumeric display module.

To overcome the above problems a third circuit technique has been evolved for use with LED and LCC displays. This technique has the following benefits:

(i) the circuit for each character is modular as with the static system

(ii) only one character generator is required for any number of characters as with the time shared system

(iii) one IC type can be used for any type of display font and current drive.

(iv) the system is static so the operating speed of LCC's is not a problem continuous a.c. drive is, however, practical to increase life of the LCC's.

(v) the number of connections to each character is minimal.

(vi) the number of connections from the character generator and clock control circuits are, therefore, minimal resulting in packaging and yield cost-benefits.

The concept is simply to use an MOS serial-in/parallel-out shift register with output circuits large enough to drive an LED direct. The output resistance of the MOS can be designed to provide current limiting for the LED and outputs can be paralleled to provide more current if required. LCC displays can be driven from the same output.

By choosing 18 bits as the basic unit a shift-register IC can be developed such that two chips can drive a 5 x 7 dot matrix alphanumeric and one chip with its outputs paralleled in pairs can drive a 7 or 9 bar numeric. The IC can also be used for other applications as discussed in Section 3 below, thus its versatility makes it economic to have a custom built IC developed. In use each character with its IC or IC's is completely modular and apart from supplies the only connections are a single input and output and a clock. Any number of characters can be strung in series and mixed LED and LCC displays are quite practical. Figure 3 shows a sketch of an 11 character alphanumeric word module.

3.   SOLID STATE DIGITAL METER WITH "ANALOGUE" READOUT

The solid state digital panel meter with LED or LCC 7-segment numeric readout offers many advantages over the traditional analogue moving coil meter, such as:

3.1.   Greater accuracy of the digital system and better stability of this accuracy.

3.2.   Better performance under various environmental conditions, e.g., vibration, temperature change, etc.

There are applications, however, where a numeric readout is not the best

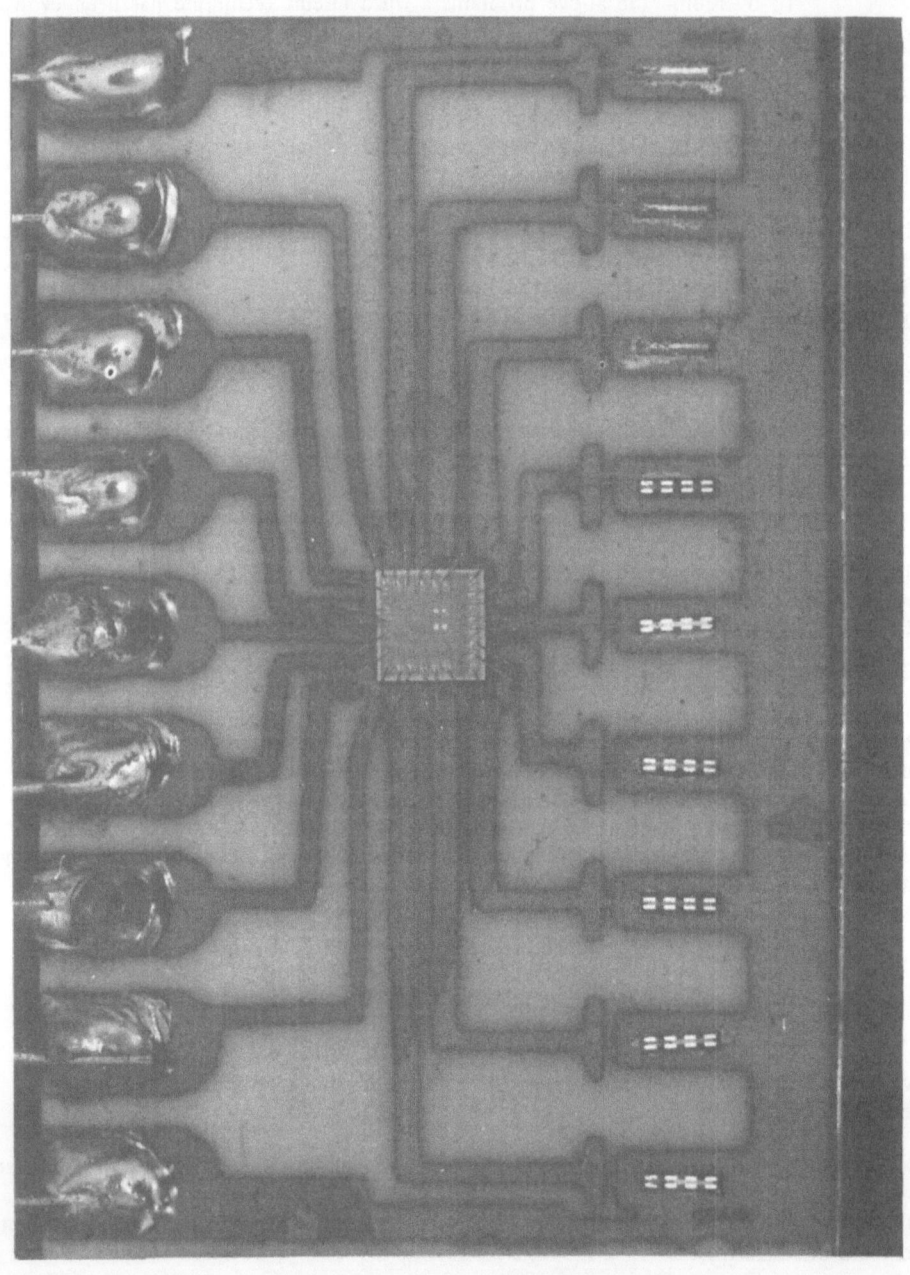

Fig. 4. "Analogue" readout module.

form of display. For instance, on the control panels of process control installations it is common to have banks of edgewise moving coil meters to display a range of variables. The operator can at a glance see the "expression on the face of the plant" whereas a bank of numbers would be difficult to comprehend. On the other hand when it comes to taking actual readings of the variables a digital display would be preferred.

A solution to the problem has been evolved which retains the digital system but arranges the LED or LCC display in a column of dots rather than in the form of numbers. At a glance the display appears like an analogue readout but it still has all the other benefits of a digital solid state meter.

The basic system consists of an analogue to digital converter-as in any other digital meter, control circuits, and a display circuit.

The display circuit consists of the same MOS serial-in/parallel-out shift register as in Section 2 above with one "bit" (memory cell) or paralleled pair of bits, for each dot in the display.

The control circuits are arranged to feed this shift register with an appropriate pulse train and clock sequence to produce the right indication. In a typical example the "writing" process may take 3 ms followed by a display period of 30 ms. Thus the input voltage is sampled and displayed 30 times per second. The display can be in the form of a "column-of-dots" or a "single-dot" display to simulate the two basic types of analogue meter display.

As with the numeric and alphanumeric displays the circuit arrangement is such that apart from power supply wires only a "data" and a "clock" wire are required between the shift register display unit and the rest of the circuit.

This unit can be made low profile and hence can be mounted flush on the panel, thus avoiding costly cut outs in the panel. The photograph in Figure 4 shows a basic module used to build up a complete display.

As the meters are normally used in groups, as explained above, the "analogue to digital" converter and control circuits can be time-shared among a number of display modules making use of the built in memory of the shift register to still retain the benefits of a flicker free static display.

4.   SUMMARY AND CONCLUSIONS

From a review of the basic requirements of display modules using LED or LCC display devices the weaknesses in traditional circuit techniques have been highlighted. A circuit technique has been proposed as an alternative which is now practical and economic using MOS IC techniques. The ability to use the same IC for numeric, alphanumeric and linear arrays of both LED and LCC has been explained. The technique offers good prospects of becoming the most economical solution to LED and LCC driving for many applications.

INDEX